GEM GUIDES

General Editor
Derek Avery

COLLINS
London and Glasgow

First published 1987
Wm Collins Sons and Co Ltd
Reprint 10 9 8 7 6 5 4 3 2 1 0

Produced by Chevprime Ltd
147 Cleveland Street
London W1

ISBN 0 00 458855

Artwork by R.P. Gossop & Co
Filmset by Computerised Typesetting Services Ltd
Printed in Great Britain by Collins, Glasgow

Contents

Amphibious Assault Ship USS *Saipan* in formation with Guided Missile Destroyer USS *Semmes* with an Adams Class ship in the background on 30 September 1980, during NATO maritime exercises.

Introduction

The 1980s find the navies of the world in considerable flux. Inflation has played a decisive part in reducing construction programmes at a time when theoretical developments have combined with such practical experiences as the Falklands campaign of 1982 to persuade maritime powers of the vital contribution that can be made by effective navies.

The result has been a resurgence in building and conversion programmes: older vessels are being updated with newer sensors and weapons; the very nature of the weapons fit has often been altered radically; and newer vessels are being produced against more stringent operational requirements. The new ships have greater flexibility, and smaller craft are being built with a decisive offensive punch in the form of several potent anti-ship missiles. Although the major powers are still producing single-role ships, these must now also possess some capac-

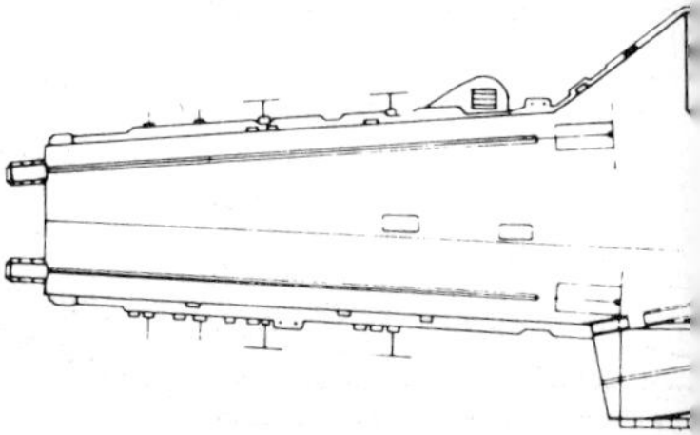

USS *Forrestal*

ity for secondary and even tertiary roles. Much the most important recent development however has been the universal acceptance of the Soviet doctrine of developing numerous multi-role types.

With the end of World War II in 1945, the future of the warship appeared in doubt. Jet-propelled, long-range aircraft and the atomic bomb seemed to herald the end of the surface ship. In essence the turbojet engine, nuclear energy, automation and electronics have contributed not to end but to alter profoundly the very nature and appearance of the modern navy. Immediately after the war, US plans for new aircraft carriers were scrapped in favour of long range strategic bombers, but the Korean War (1950-53), together with the first nuclear strike aircraft (1951), made US politicians increasingly aware of the necessity of a US-controlled forward-based nuclear strike force; thus the first new postwar carrier, the USS *Forrestal*,

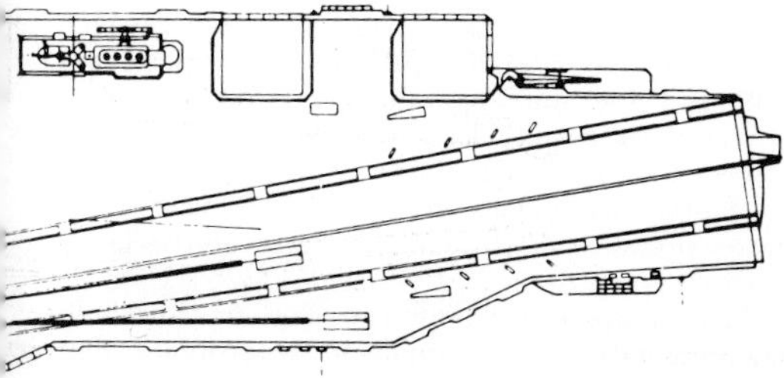

was completed in 1955. The USS *Forrestal* incorporated an angled flight deck, a British innovation, and a complement of twin-jet Douglas A3D Skywarrior aircraft.

US carrier production continued with a total of eight oil-fuelled craft being built. In addition to these, five nuclear-powered carriers were built, including the four 94,000-ton Nimitz class super-carriers, the largest and most expensive ships ever built, of which the latest, *Theodore Roosevelt,* was commissioned on 29 October 1986.

Carrier production elsewhere has not been significant. France commissioned *Clemenceau* and *Foch* in the early 1960s, Italy *Giuseppe Garibaldi* in 1985, whilst the USSR have to date only managed the four Kiev class ships, although larger craft have been laid down. Great Britain, meanwhile, can currently boast three Invincible class carriers with their complements of Sea Harriers and Sea King helicopters. It was Britain, and not the United States, which produced the three features essential to the jet-age carrier. These were the angled flight deck, the steam catapult, and the mirror landing system, which all appeared in the early 1950s. Today only the two super powers are able to afford a large modern fleet with everything that entails in quality and quantity of ships, aircraft, people and the necessary back-up facilities.

The development of guided missiles – first installed in existing warships rebuilt to incorporate the

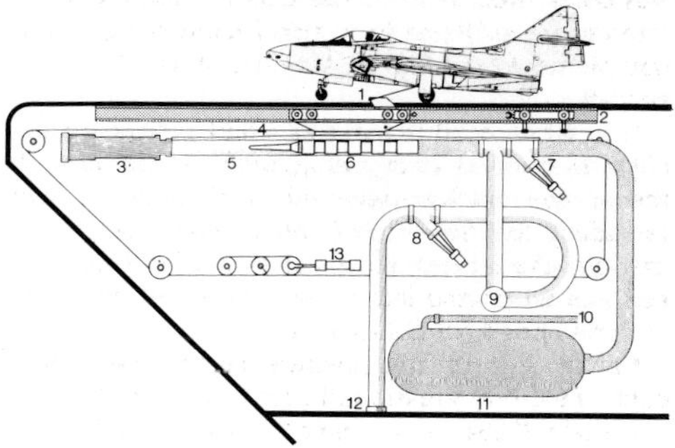

Operation of the Steam Catapult

1 Towing strop
2 Shuttle return grab
3 Retardation cylinder
4 Cylinder seal
5 Twin cylinder tubes
6 Piston and shuttle assembly
7 Launch valve
8 Exhaust valve
9 Exhaust collector box
10 High pressure steam supply

launchers, magazines, radars and control centres, often at the expense of all other armament – led to new purpose-built missile ships, first initiated by the US Navy and followed by the British and Soviet navies. Along with this, more advanced radars were developed for the detection of targets and for missile guidance, together with more computers with enhanced abilities to plot and predict interception points. But as one weapon is born, it is followed by its counter weapons and countermeasures.

Today, most US cruisers and destroyers are fitted with Harpoon anti-ship missiles with a range of 97km (60mls), plus, in some instances, Tomahawk anti-ship or land-attack cruise missiles. Warships generally are being fitted with 'paint-defence' or 'close-in' weapons (multi-barrel cannon such as Gatling guns) and advanced missile fire-control systems, capable of simultaneous engagement of several targets.

The most modern of all missile-control systems is employed in *Ticonderoga,* the latest US missile cruiser. The installation of her Aegis weapons-control system activates an autopilot guidance system in each missile at the time of launch. At the same time, *Ticonderogas*'s SPY-1 phased-array radar continuously tracks numerous targets and missiles in flight, adjusting the missile autopilot as necessary in flight to guide them near their targets, the detailed radar guidance only being necessary for the few seconds before contact. The *Ticonderoga* has four guidance systems, each capable of five simultaneous oper-

GENERAL DYNAMICS BGM-109B TOMAHAWK

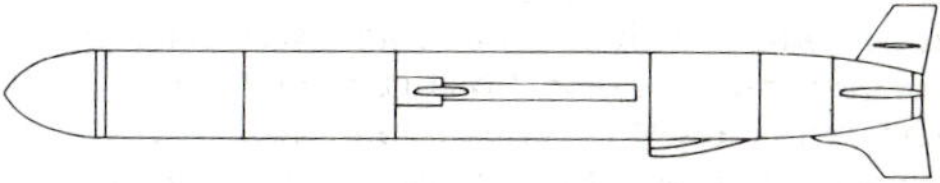

Country of Origin: USA

Type: Submarine and ship-launched anti-ship cruise missile

Dimensions: Length 6.40m (21ft 0in); span 2.54m (8ft 4in); diameter 53.0cm (20.86in)

Weights: Total round 1,202kg (2,650lb); warhead 454kg (1,000lb) high explosive

Guidance: For cruise, a Lear-Siegler or Northrop strapdown inertial platform and a Honeywell AN/APN-194 radar altimeter operating with an IBM 4Pi SP-OA computer, and for attack a Texas Instruments PR53/DSQ-28 two-axis active radar seeker

Control: Cruciform rear fins

Propulsion: For boost, one Atlantic Research solid-propellant rocket delivering a thrust of 3,175kg (7,000lb) for 7 seconds, and for sustain one Williams Research F107-WR-100 turbofan delivering a thrust of 272kg (600lb)

Performance: Maximum speed 885km/h (550mph) at sea level; range 450km (280mls)

ations, making her ten times as effective as conventional ships. The Aegis and SPY-1 systems are backed up by huge ultra-fast computers interpreting the data from the phased-array radar from which continual updating is made to the missile autopilot. Phased-array radars have fixed antennae that search in different directions with numerous pencil-thin beams from each fixed antenna array, known as a 'face'. These beams are computer-steered and capable of searching areas much faster than the continual rotating radars. In addition they are much more difficult to jam and have the ability to switch rapidly from search to guidance mode.

The nearest counterpart technologically to the USS *Ticonderoga* are the Soviet nuclear Kirov class cruisers. They have phased-away missile-control radars, and the *Kirov* itself has two cone-shaped guidance radars which only afford a limited field of view, one either end of her superstructure and each capable of simultaneous guidance of six missiles. The second systems around the *Kirov* are only in their early stages of development and lack the target-handling capacity of the SPY-1 system. However, this is an area in which the Soviet Union is making big advances and the next Kirov class cruiser will undoubtedly boast a much improved system.

right: USS *Ticonderoga*

GLOSSARY

AA: Anti-aircraft
AGR: Radar picket ship
AGS: Surveying ship
ARM: Anti-radiation missile
A/S, ASW: Anti-submarine (warfare)
ASM: Air-to-surface missile
bhp: see Horsepower
Cal: Calibre (the diameter of a gun barrel)
CC: Cruiser
CG: Cruiser, guided-missile
CL: Cruiser, light
CODAG, CODOG, COGAG, COGOG, COSAG: Descriptions of mixed propulsion systems: combined diesel and gas turbine, diesel or gas turbine, gas turbine and gas turbine, gas turbine or gas turbine, steam and gas turbine
DC: Depth charge
DCT: Depth charge thrower
DD: Destroyer
DDG: Destroyer, guided-missile
Displacement: Basically the weight of water displaced by a ship's hull when floating: (a) Light – without fuel, water or ammunition; (b) Normal – used for Japanese MSA ships (similar to 'Standard'); (c) Standard – (as defined by Washington Naval Conference 1922) fully manned and stored, but without fuel or reserve feed-water; (d) Full load – fully laden with all stores, ammunition, fuel and water
DSRV: Deep-submergence recovery vessel
ECM: Electronic counter measures, e.g. jamming
ELINT: Electronic intelligence, e.g. recording radar, W/T, etc.
FF: Frigate
FFG: Frigate, guided-missile
FRAM: US Navy's 'Fleet rehabilitation and modernization' programme
GRP
Horsepower: Power developed or applied: (a) bhp – brake horsepower (power available at the crankshaft); (b) shp – shaft power (power delivered to the propeller shaft); (c) ihp – indicated horsepower (power produced by expansion of gases in the cylinders of reciprocating steam engines). N.B.: if the type of horsepower is not known, hp is used.
hp: see Horsepower
ihp: see Horsepower
Knot (kt): 1 nautical mile (1.85 km) per hour
kW: Unit of power equivalent

to 1,000 watts; 1 watt = 1.341 $\times 10^{-3}$ horsepower
LCM: Landing craft, mechanised
LCU: Landing craft, utility
Length: Expressed in various ways: (a) overall – length between extremities; (b) pp – between perpendiculars (between fore side of the stem and after side of the rudderpost); (c) wl – water-line (between extremities on the water line)
LPD: Amphibious transport dock (US); Assault ship (UK)
LSM: Landing ship, medium
LST: Landing ship, tank
MAP: Military Aid Programme (US)
MIRV: Multiple independently targetable re-entry vehicle (on ballistic missiles)
MRV: Multiple re-entry vehicle
MSC: Military Sealift Command (US)
nm: Nautical mile(s); 1nm = (1.85 km) = average meridian length of one minute of latitude
rpm: Revolutions per minute (of engines, propellers, radar aerials, etc.)
SAM: Surface-to-air missile
shp: see Horsepower
SLBM: Submarine-launched ballistic missile
SS: Attack submarine
SSB: Submarine, ballistic-missile
SSBN: Submarine, nuclear-powered ballistic-missile (French, SNLE)
SSG: Guided-missile submarine
SSGN: Nuclear-powered guided-missile submarine
SSM: Surface-to-surface missile
SSN: Nuclear-powered attack submarine
Tonnage: Weight in tons, computed on capacity of a ship's hull rather than its 'displacement': (a) Gross – the internal volume in cubic feet divided by 100 of all spaces within the hull and all permanently enclosed spaces above decks that are available for cargo, stores and accommodation; (b) Net – gross tonnage minus all those spaces used for machinery, accommodation, etc. ('non-earning' spaces); (c) Deadweight (dwt) – the amount of cargo, bunkers, stores, etc., that a ship can carry at her load draught
VSTOL: Vertical or short take-off/landing
VTOL: Vertical take-off/landing

SHIP DESIGNATIONS

AIRCRAFT CARRIERS

Attack Aircraft Carriers: (US) Nimitz and Enterprise classes

(Nuclear)

Attack Aircraft Carriers

ASW Aircraft Carriers

Light Aircraft Carriers: Invincible and Moskva classes

MAJOR SURFACE SHIPS

Cruisers: Over 10,000 tons, including missile conversions

Light Cruisers: 7,000 tons to 10,000 tons

Destroyers: 3,000 to 7,000 tons, plus original conventional destroyers

Frigates: 1,100 to 3,000 tons

Corvettes: 500 to 1,100 tons

LIGHT FORCES

Fast Attack Craft (FAC) (25 knots and above): FAC (Missile), FAC (Gun), FAC (Torpedo), FAC (Patrol)

Patrol Craft (below 25 knots): Large Patrol Craft (100 to 500 tons), Coastal Patrol Craft (below 100 tons)

AMPHIBIOUS FORCES: Command Ships, Assault Ships, Landing Ships, Landing Craft, Transports

MINE WARFARE FORCES: Minelayers, MCM Support Ships, Mine Sweepers (Ocean), Mine Hunters, Mine Sweepers (Coastal), Mine Sweepers (Inshore), Mine Sweeping Boats

SUBMARINES

Strategic Missile: Nuclear-propelled and con-

ventionally propelled
Cruise Missile: Nuclear-propelled and conventionally propelled
Fleet Submarines: Nuclear-propelled, torpedo attack
Patrol Submarines: Conventionally propelled, torpedo attack

Notes about the text

Complements are identified as follows: first figure, ships officers; second figure, other ratings; third figure, air-wing complement.

The number of ships in any class are identified by country, first by the number of ships laid down followed by the number of ships still to be commissioned or undergoing refits, and finally ships ordered but not yet laid down.

Due to limitations of space, where a class is too numerous only the first of the class has been listed.

CLEMENCEAU CLASS
AIRCRAFT CARRIER
Country of Origin: France

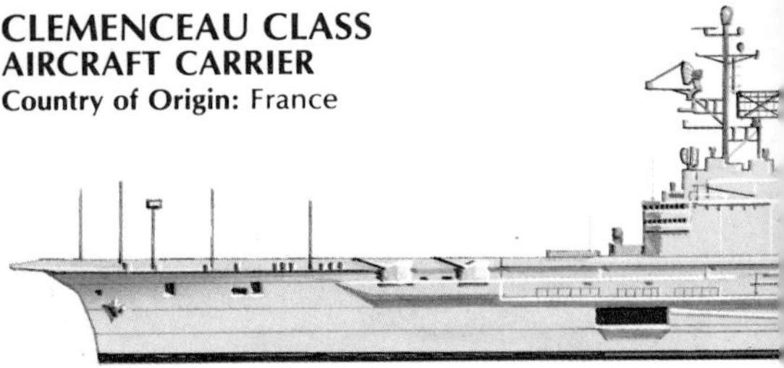

Displacement: 27,310 tons normal and 32,780 tons full load
Dimensions: Length 265.0m (869.4ft); beam 31.72m (104.1ft); draught 8.6m (28.2ft); flight deck length 257.0m (843.2ft) and width 29.5m (96.8ft)
Gun Armament: Eight 100mm (3.9in) L/55 DP in single mountings
Aircraft: 40 fixed-wing (20 Dassault-Breguet Super Etendards, 10 Vought F-8 Crusaders and 10 Dassault-Breguet Alizes) and four rotary-wing (two Super Frelons and two Alouette IIIs)
Electronics: One Thompson-CSF DRBV 20 surveillance radar, one Thompson-CSF DRBC 20C air-warning radar, two Thompson-CSF DRBI 10 height-finding radars, two Thompson-CSF DRBC 32 fire-control radars, one Decca 1226 navigation radar, one NRBA landing radar, one SQS-503 sonar, one SENIT 2 combat information system, and comprehensive electronic countermeasures

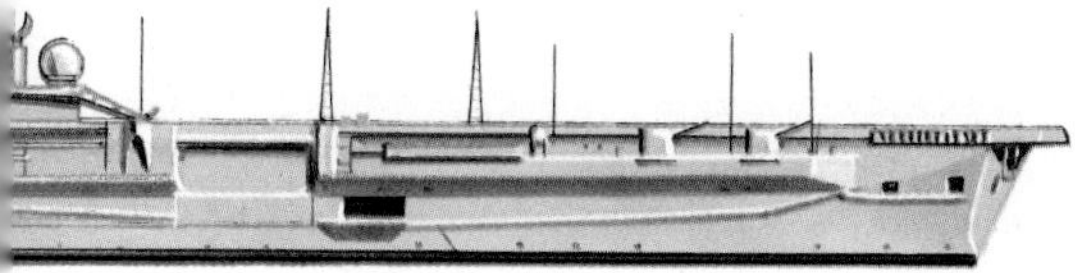

Propulsion: Six boilers supplying steam to two sets of Parsons geared turbines delivering 93,960kW (126,000shp) to two shafts
Performance: Maximum speed 32kts; range 13,900km (8,635mls) at 18kts or 6,500km (4,040mls) at 32kts
Complement: 64 + 476 + 798
Class
1. France (2)
CLEMENCEAU Commissioned November 1961
FOCH Commissioned July 1963
Remarks: Apart from the USA and USSR, France is now the only country in the world to operate large aircraft carriers for a multi-role air group of fixed-wing aircraft, complemented by helicopters. The design features an angled flight deck, two aircraft lifts, two steam catapults and two mirror landing aids. Both ships have been modernized and refitted, and it is planned to keep the *Clemenceau* in service until 1992 and the *Foch* until 1998.

ENTERPRISE CLASS NUCLEAR-POWERED MULTI-ROLE AIRCRAFT CARRIER

Country of Origin: USA

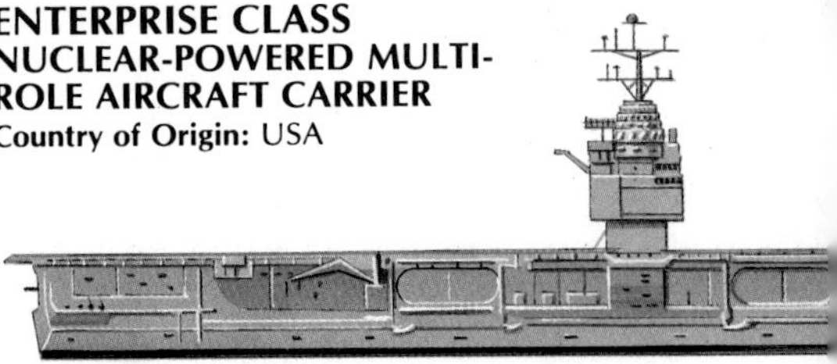

Displacement: 75,700 tons standard and 89,600 tons full load
Dimensions: Length 335.9m (1,102ft); beam 40.5m (133ft); draught 10.9m (35.8ft); flight deck width 76.8m (252ft)
Gun Armament: Three Phalanx Mk 15 20mm (0.8in) close-in weapon systems, and three 20mm (0.8in) Mk 68 mountings
Missile Armament: Three Mk 57 launchers for RIM-7 NATO Sea Sparrow surface-to-air missiles
Aircraft: About 84
Electronics: One Raytheon/Sylvania SPS-10 surface-search radar, one ITT/Gilfillan SPS-48C 3D air-search radar, one Raytheon SPS-49 air-search radar, one Westinghouse SPS-65 combined-search radar, one Westinghouse SPS-58 low-level air-search radar, three SPN series aircraft-landing radars, three Raytheon Mk 91 missile fire-control systems, Naval Tactical Data system, OE-82 satellite communications

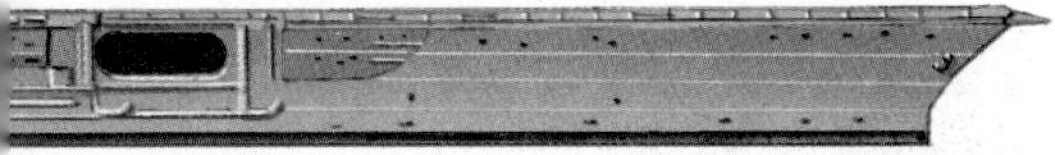

antenna, URN-26 TACAN, four Mk 36 Super Chaffroc rapid-blooming overhead chaff systems, and one SLQ-ESM suite

Propulsion: Eight pressurized water-cooled Westinghouse A2W reactors supplying steam to four sets of Westinghouse geared turbines delivering 208,795kW (280,000shp) to four shafts

Performance: Maximum speed 35kts; range 741,245km (460,600mls)

Complement: 162 + 2,940, plus a carrier air-wing complement of 304 + 2,323

Class

1. USA (1)

ENTERPRISE Commissioned November 1961

Remarks: The *Enterprise* (CVN 65) was the world's second nuclear-powered surface warship, and the design was based on that of the highly successful Forrestal class. She is to be extensively upgraded in 1993-95.

FORRESTAL CLASS
MULTI-ROLE AIRCRAFT CARRIER

Country of Origin: USA

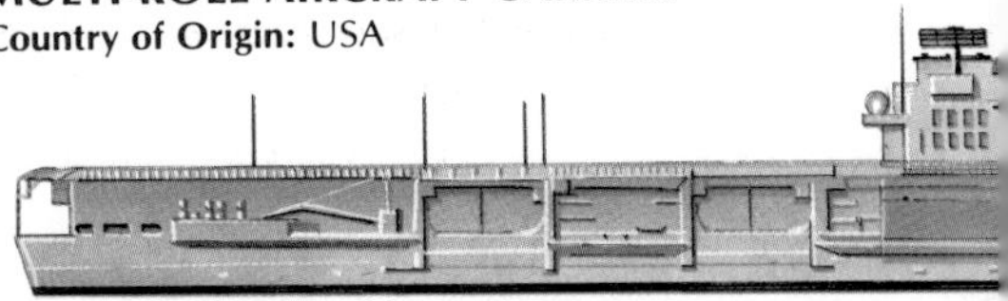

Displacement; 56,060 tons standard and 75,900 tons full load
Dimensions: Length 331.0m (1,086ft); beam 39.5m (129.5ft); draught 11.3m (37ft); flight deck width 76.8m (252ft)
Gun Armament: To be fitted with three 20mm (0.8in) Phalanx Mk 16 close-in weapon system mountings
Missile Armament: Two Mk 25 octuple launchers for RIM-7 Sea Sparrow surface-to-air missiles
Aircraft: About 70
Electronics: One Westinghouse SPS-58 low-angled air-search radar, one Raytheon/Sylvania SPS-10 surface-search radar, one ITT/Gilfillan SPS-48 3D air-search radar, one Raytheon SPS-49 air-search radar, one LN-66 navigation radar, two Mk 115 missile fire-control systems, Naval Tactical Data System, URN-20 TACAN, OE-82 satellite communications antenna, three Mk 36 Super Chaff launchers, and one SLQ-29 ESM suite
Propulsion: Eight Babcock & Wilcox boilers supplying steam to four sets of Westinghouse geared turbines

delivering 193,880kW (260,000shp)
Performance: Maximum speed 33kts; range 14,805km (9,200mls) at 20kts or 7,400km (4,600mls) at 30kts

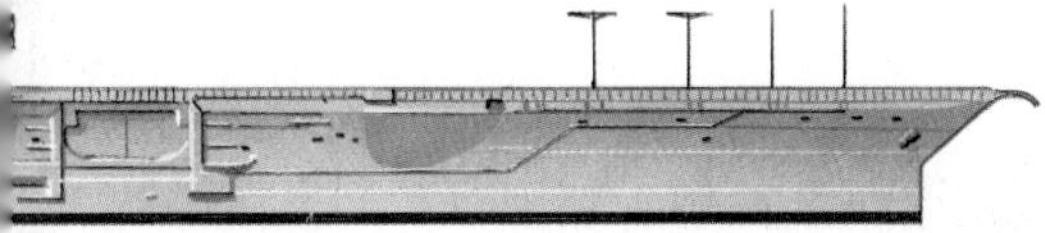

Complement: 145 + 2,645, plus a carrier air-wing complement of about 2,150
Class
1. USA (4)
FORRESTAL Commissioned October 1955
SARATOGA Commissioned April 1956
RANGER Commissioned August 1957
INDEPENDENCE Commissioned January 1959
Remarks: The four units of the Forrestal class were the first aircraft carriers built specifically with jet-aircraft operations in mind, and their design was delayed by the decision to incorporate the British-designed angled flight deck and steam catapult. Other notable features are the enclosed bow (the first on American carriers) for greater seaworthiness, an armoured flight deck, massive underwater protection, and internal compartmentation design to mitigate the effects of nuclear as well as conventional damage. Apart from the four catapults and its angled flight deck, each ship has four deck-edge lifts (three to starboard and one to port).

GARIBALDI CLASS
LIGHT AIRCRAFT CARRIER

Country of Origin: Italy

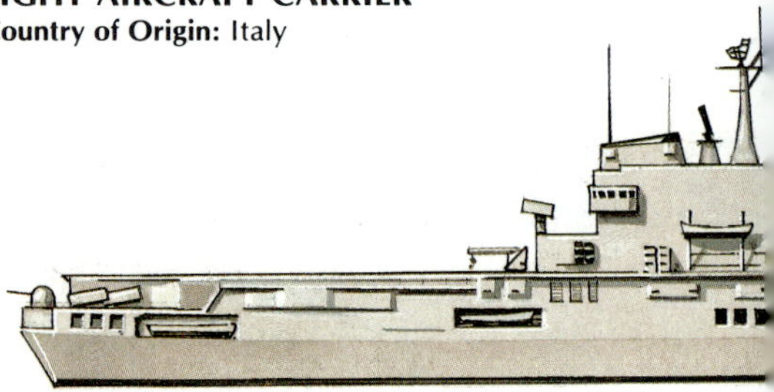

Displacement: 10,100 tons standard and 13,370 tons full load

Dimensions: Length 180.0m (590.4ft); beam 30.4m (99.7ft); draught 6.7m (22ft); flight-deck length 174.0m (570.7ft) and width 30.4m (99.7ft)

Gun Armament: Six 40mm Breda L/70 AA in three twin mountings

Missile Armament: Two Twin Teseo launchers for 10 Otomat Mk 2 surface-to-surface missiles, and two Albatros launchers for Aspide surface-to-air missiles

Anti-submarine Armament: Two triple ILAS 3 tube mountings for 324mm (12.75in) A244S or 324mm (12.75in) Mk 46 A6S torpedoes, and helicopter-launched weapons

Aircraft: Up to 18 Agusta-Sikorsky SH-3 Sea King helicopters

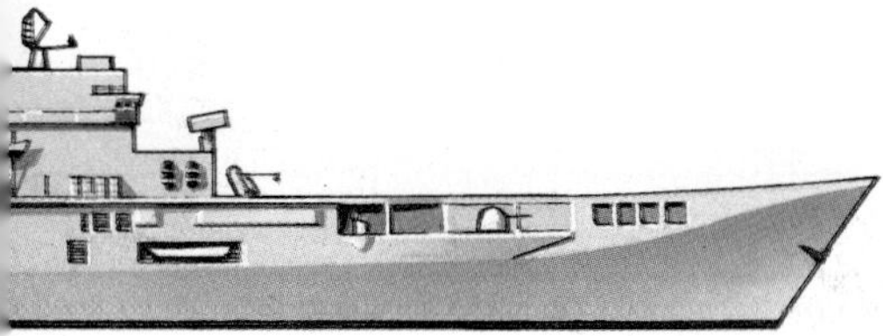

Propulsion: Four General Electric-Fiat LM2500 gas turbines delivering 59,655kW (80,000hp) to two shafts
Performance: Maximum speed 30kts; range 13,000km (8,080mls) at 20kts
Complement: 550 (accommodation is available for 825)
Class
1. Italy (1)
GIUSEPPE GARIBALDI Commissioned 1985
Remarks: Designed to replace the *Andrea Doria* and *Caio Duilio,* the *Giuseppe Garibaldi* features extremely comprehensive offensive and defensive missile fits, excellent sensor fit and substantial anti-submarine capability.

INVINCIBLE CLASS
LIGHT AIRCRAFT CARRIER

Country of Origin: UK

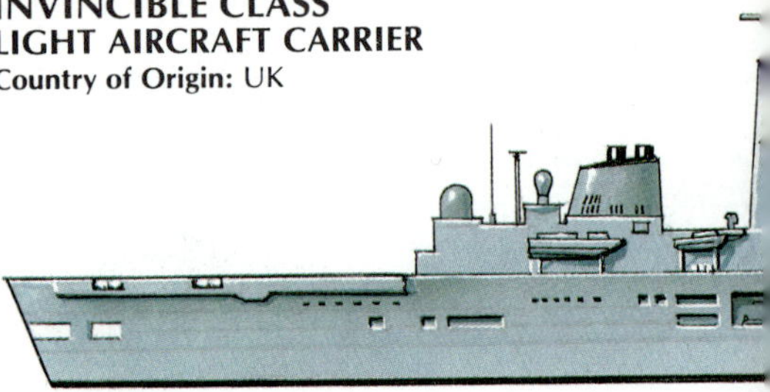

Displacement: 16,000 tons standard and 19,500 tons full load
Dimensions: Length 206.6m (677ft); beam 27.5m (90ft); draught 7.3m (24ft); flight-deck length 167.8m (550ft) and width 31.9m (105ft)
Gun Armament: Two 20mm (0.8in) Phalanx CIWS mountings and two 20mm (0.8in) AA
Missile Armament: One twin launcher for 22 Sea Dart surface-to-air missiles
Anti-submarine Armament: Helicopter-launched weapons
Aircraft: Five BAe Sea Harrier FRS.Mk 1 V/STOL aircraft and nine Westland Sea King HAS.Mk 2/5 helicopters
Propulsion: Four Rolls-Royce Olympus TM3B gas turbines delivering 83,520kW (112,000shp) to two shafts

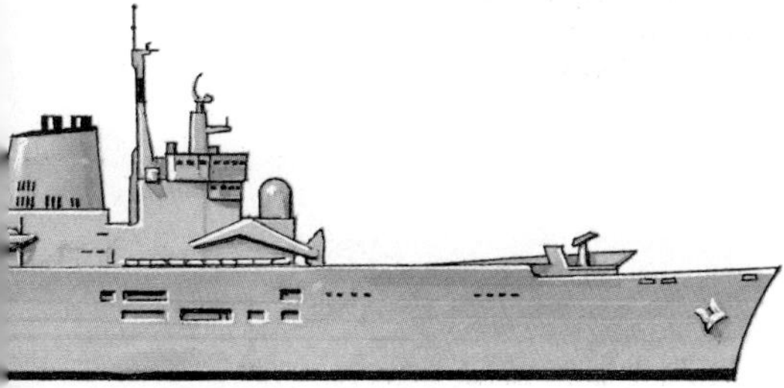

Performance: Maximum speed 28kts; range 9,250km (5,750mls) at 15kts
Complement: 131 + 131 + 869 plus 320 air group personnel
Class
1. UK (3)
INVINCIBLE Commissioned July 1980
ILLUSTRIOUS Commissioned July 1982
ARK ROYAL Commissioned November 1984
Remarks: The class has provision for a revived fixed-wing strength, in the form of BAe Sea Harrier multi-role aircraft. These can be fitted onto the existing pair of lifts, and are provided with take-off aid by the provision of a 'ski-jump' lip to the flight deck (7° on the first two and 15° on the last). Point-defence weapons have also been increased considerably as a result of experience in the Falklands.

JOHN F. KENNEDY CLASS
MULTI-ROLE AIRCRAFT CARRIER

Country of Origin: USA

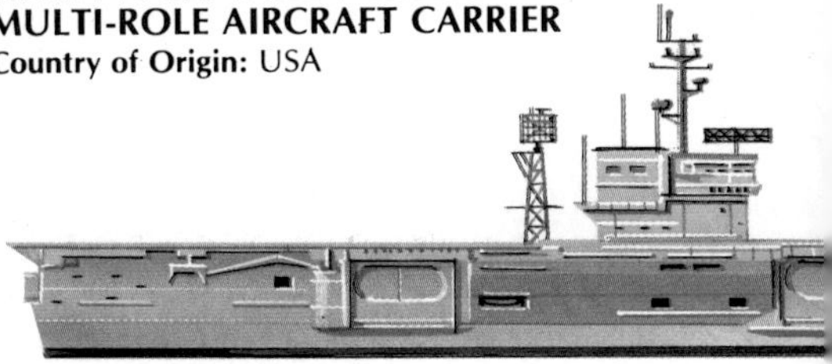

Displacement: 61,000 tons standard and 82,000 tons full load
Dimensions: Length 320.7m (1,052ft); beam 39.6m (130ft); draught 10.9m (35.9ft); flight deck width 76.9m (252ft)
Gun Armament: Three 20mm (0.8in) Phalanx Mk 15 close-in weapon system mountings
Missile Armament: Three Mk 29 launchers for RIM-7 Sea Sparrow surface-to-air missiles
Aircraft: About 85
Electronics: One Raytheon/Sylvania SPS-10 surface-search radar, one Hughes/Westinghouse SPS-43 air-search radar, one Raytheon SPS-49 air-search radar, one Westinghouse SPS-58 air-search radar, one SPN-10 navigation radar, two SPN-42 landing radars, three Mk91 missile fire-control systems, URN-20

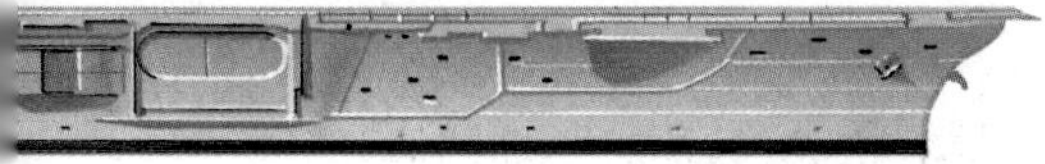

TACAN, Naval Tactical Data System, several advanced electronic warfare systems, OE-82 satellite communications antenna, four Mk 36 Super Chaffroc rapid-blooming overhead chaff launchers, and one SLQ-29 ESM suite

Propulsion: Eight Foster-Wheeler boilers supplying steam to four sets of Westinghouse geared turbines delivering 208,795kW (280,000shp) to four shafts

Performance: Maximum speed 30+kts; range 14,805km (9,200mls) at 20kts

Complement: 150 + 2,645, plus a carrier air wing complement of about 2,150

Class

1. USA (1)

JOHN F. KENNEDY Commissioned September 1968

Remarks: The *John F. Kennedy* is similar in all essential respects to the Kitty Hawk class, and was originally considered for nuclear propulsion.

KIEV CLASS AIRCRAFT CARRIER

Country of Origin: USSR
Displacement: 36,000 tons standard and 42,000 tons full load

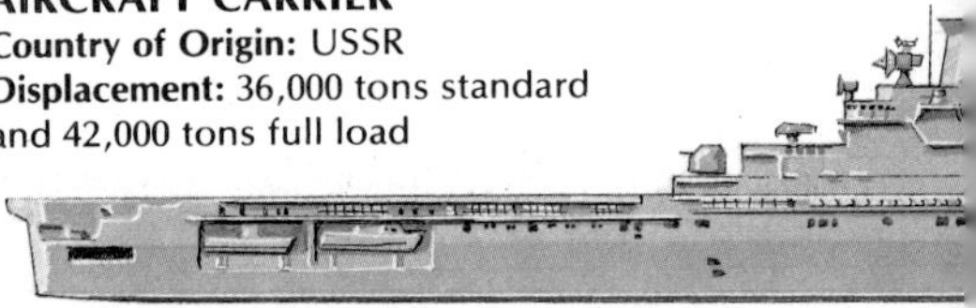

Dimensions: Length 274.0m (898.7ft); beam 41.2m (135ft); draught 10.0m (32.8ft); overall width 50.0m (164ft) including flight deck and sponsons
Gun Armament: Four 76mm (3in) L/60 DP in two mountings, and eight 30mm (1.2in) AA 'Gatling' mountings
Missile Armament: Four twin launchers for 16 SS-N-12 surface-to-surface missiles, two twin launchers for 72 SA-N-3 'Goblet' surface-to-air missiles, and two twin launchers for about 36 SA-N-4 surface-to-air missiles
Torpedo Armament: Two quintuple 533mm (21in) tube mountings
Anti-submarine Armament: One twin SUW-N-1 launcher for 20 FRAS-1/SS-N-14 'Silex' missiles, two RBU 6,000 12-barrel rocket launchers, and helicopter-launched weapons
Aircraft: Up to 32 are normally carried, a typical mixture being 12 Yakovlev Yak-36 'Forger-A' and one Yak-36 'Forger-B' fixed-wing aircraft, and 16 Karmov Ka-25 'Hormone-A' and three Ka-25 'Hormone-B' helicopters

Propulsion: Boilers supplying steam to four sets of geared turbines delivering 134,225kW (180,000shp) to four shafts
Performance: Maximum speed 32kts; range 24,100km (14,975mls) at 18kts or 7,400km (4,600mls) at 30kts
Complement: 2,500 including air group personnel
Class
1. USSR (3+1)
KIEV Commissioned May 1975
Remarks: Each ship is able to carry a useful number of fixed and rotary-wing aircraft for launch and recovery from the slightly angled flight deck. Two lifts are provided for movement of aircraft between the flight deck and hangar. Apart from its air component (whose fixed-wing Yakovlev Yak-36 'Forger' VTOL aircraft are of uncertain capability), each Kiev-class ship carries a prodigious anti-ship and anti-aircraft missile armament, together with effective anti-air point-defence gun armament and limited anti-submarine capability with rocket launchers.

KITTY HAWK CLASS
MULTI-ROLE AIRCRAFT-CARRIER

Country of Origin: USA

Displacement: 60,100 tons standard and 80,800 tons full load for *Kitty Hawk* and *Constellation;* D60,300 tons standard and 78,500 tons full load for *America*
Dimensions: Length 318.8m (1,064ft) for *Kitty Hawk* and *Constellation* and 319.3m (1,04.57ft) for *America;* beam 39.6m (130ft); draught 11.3m (37ft); flight deck width 76.9m (252ft)
Gun Armament: Three 20mm Phalanx Mk 15 close-in weapon system mountings

Missile Armament: Three Mk 29 octuple launchers for 24 RIM-7 Sea Sparrow surface-to-air missiles
Anti-submarine Armament: Air-dropped weapons
Aircraft: About 85
Electronics: One Hughes SPS-52 radar, one Hughes/ Westinghouse SPS-43 air-search radar, one General Electric SPS-30 height-finding radar, one SPN-10 navigation radar, one SPN-42 navigation radar, two Mk 91 missile fire control systems, URN-20 TACAN, Naval Tactical Data System, several advanced electronic warfare systems, OE-82 satellite communications

antenna, four Mk 36 Super Chaffroc rapid-blooming overhead chaff launchers, and one SLQ-29 ESM suite
Propulsion: Eight Foster-Wheeler boilers supplying steam to four sets of Westinghouse geared turbines

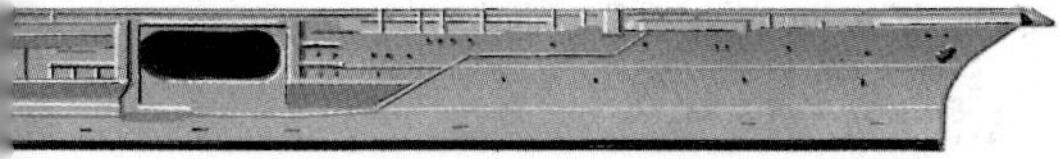

delivering 208,795kW (280,000shp) to four shafts
Performance: Maximum speed 30+kts; range 14,805km (9,200mls) at 20kts or 7,400km (4,600mls) at 30kts
Complement: 150+2,645, plus a carrier air wing complement of about 2,150
Class
1. USA(3)
KITTY HAWK Commissioned April 1961
CONSTELLATION Commissioned October 1961
AMERICA Commissioned January 1965
Remarks: Designed as improved 'Forrestal' class carriers, the 'Kitty Hawk' and single 'John F Kennedy' class ships were the US Navy's last fossil-fuelled carriers to be built. There are two catapults each on the bow and the angled flight deck, and four deck-edge lifts (one port and three starboard). Three 20mm (0.8in) Mk 16 Phalanx CIWS mountings are being fitted on each ship, and service life extension program refits are planned, starting in 1987-89.

MIDWAY CLASS
MULTI-ROLE AIRCRAFT-CARRIER

Country of Origin: USA

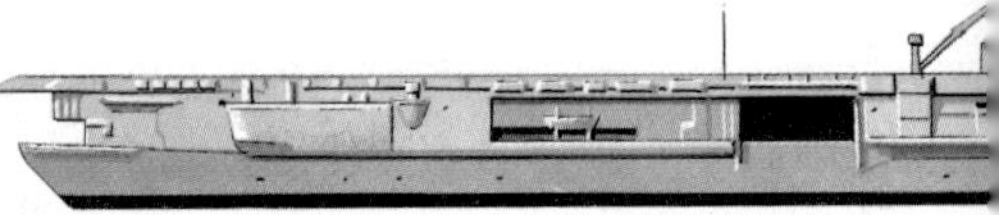

Displacement: 51,000 tons standard and 62,200 tons full load for Cv 41, or 52,500 tons standard and 62,200 tons full load for CV 43
Dimensions: Length 298.4m (979ft); beam 36.9m (121ft); draught 10.8m (35.3ft); flight-deck width 72.5m (238ft)
Gun Armament: Three 20mm (0.8in) Phalanx Mk 15 close-in weapon system mountings
Missile Armament: Two Mk 25 launchers for RIM-7 Sea Sparrow surface-to-air missiles
Aircraft: About 75
Armour: Multi-layer protection afforded by thin armour on several decks
Electronics: One ITT/Gilfillan SPS-48C 3D radar, one Westinghouse SPS-65V combined air and surface-search radar, one Hughes/Westinghouse SPS-43C air-search radar, one SPS-49, one LN-66 navigation radar, one SPN-35A carrier-controlled approach radar, one SPN-42 carrier controlled approach radar, one SPN-44 carrier-controlled approach radar, two Mk

115 missile fire-control systems, URN-20 TACAN,

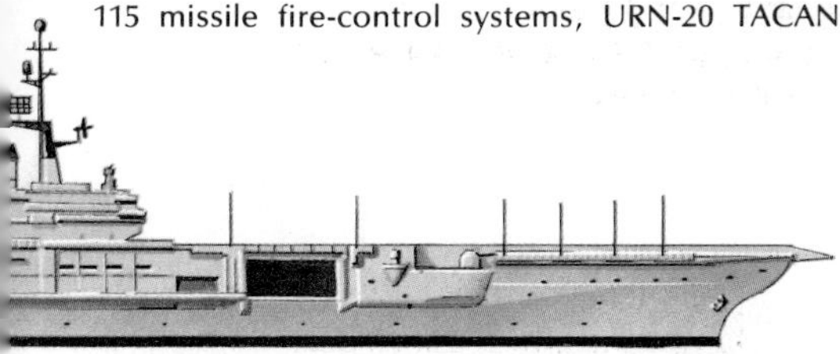

Naval Tactical Data System, OE-82 satellite communications antenna, one SLQ-29 ESM suite, and four Mk 36 Super Chaffroc rapid-blooming overhead chaff launchers
Propulsion: 12 Babcock & Wilcox boilers supplying steam to four sets of Westinghouse geared turbines delivering 158,090kW (212,000shp) to four shafts
Performance: Maximum speed 30+kts
Complement: 140+2,475 plus an air wing strength of about 1,800
Class
1. USA(2)
MIDWAY Commissioned September 1945
CORAL SEA Commissioned October 1947
Remarks: The oldest first-line carriers left in US Navy service, the two 'Midway' class multi-role carriers are survivors of the building programme instituted in World War II. Their size precludes the carriage of a full carrier air wing, however, and the two ships are thus allocated to slightly less significant roles.

NIMITZ CLASS
NUCLEAR-POWERED MULTI-ROLE AIRCRAFT-CARRIER

Country of Origin: USA
Displacement: 72,700 tons light, 81,600 tons standard and 91,485 tons full load (CVNs 68-70) or 96,350 tons full load (CVNs 71-73)
Dimensions: Length 332.9m (1,092ft); beam 40.8m (134ft); draught 11.3m (37ft); flight-deck width 76.8m (252ft)
Gun Armament: Two 20mm (0.8in) Phalanx Mk 15 close-in weapon system mountings
Missile Armament: Three Mk 29 launchers for 24 RIM-7 Sea Sparrow surface-to-air missiles
Aircraft: 90+
Propulsion: Two pressurized-water cooled nuclear reactors (Westinghouse A4W or General Electric A1G) supplying steam to four sets of geared turbines delivering 193,880kW (260,000shp) to four shafts
Performance: Maximum speed 30+kts
Complement: 3,300, and a carrier air wing strength of 3,000

Class
1. USA(3+3+2)
NIMITZ Commissioned May 1975

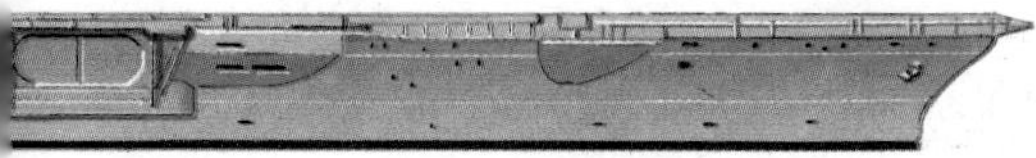

DWIGHT D EISENHOWER Commissioned October 1977
CARL VINSON Commissioned February 1982
THEODORE ROOSEVELT Commissioned 1987
ABRAHAM LINCOLN Commissioned 1990
GEORGE WASHINGTON Commissioned 1990
Remarks: Undoubtedly the most powerful surface warships in the world, the planned eight units of the 'Nimitz' class will provide the US Navy with the core of its carrier task group strength from the 1990s onwards. Together with nuclear-powered escorts, these magnificent ships offer the possibility of high-speed long range operations limited only by the aviation stores (fuel and munitions) and food that can be supplied by underway replenishment ships. The normal fit of four catapults and four deck-edge lifts is retained, and Phalanx CIWS mountings are to be added.

IOWA CLASS
BATTLESHIP

Country of Origin: USA
Displacement: 45,000 tons standard and 58,000 tons full load

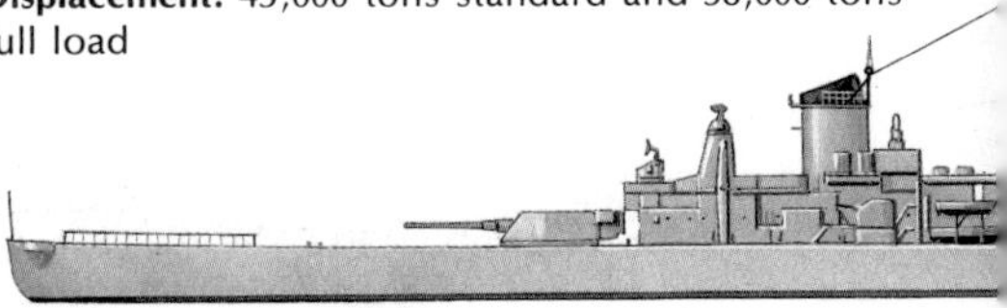

Dimensions: Length 270.4m (887.2ft); beam 33.0m (108.2ft); draught 11.6m (38ft)
Gun Armament: Nine 406mm (16in) L/50 in three triple mountings, 12 127mm (5in) L/38 DP in six Mk 38 twin mountings, and four 20mm (0.8in) Phalanx Mk 15 close-in weapon system mountings
Missile Armament: Eight quadruple launchers for 32 BGM-109 Tomahawk cruise missiles, and four launchers for 16 RGM-84A Harpoon surface-to-surface missiles
Anti-submarine Armament: Helicopter-launched weapons
Aircraft: Three or four helicopters on a platform aft
Armour: 343/41mm (13.5/1.62in) belt, 152mm (6in) decks, 432/184mm (17/7.25in) turrets, 439mm (17.3in) barbettes, and 445/184mm (17.5/7.25in) conning tower
Propulsion: Eight Babcock & Wilcox boilers supplying steam to four sets of geared turbines (General Electric

in BBs 61 and 63, Westinghouse in BBs 62 and 64) delivering 158,090kW (212,000shp) to four shafts

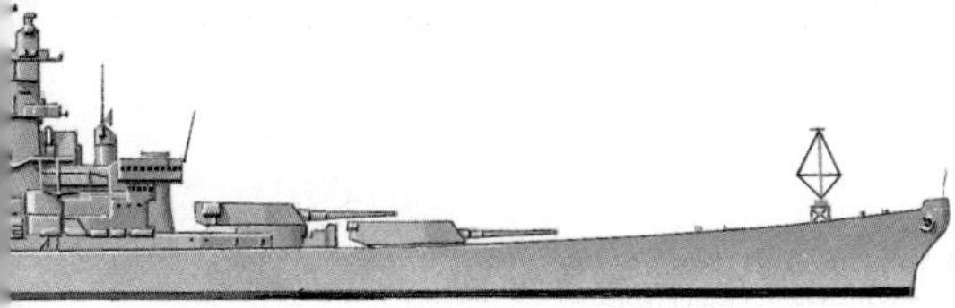

Performance: Maximum speed 33kts; range 27,800km (17,275mls) at 17kts
Complement: 1,571
Class
1. USA (2+2)
IOWA Commissioned February 1943
NEW JERSEY Commissioned May 1943
MISSOURI Commissioned June 1944
WISCONSIN Commissioned April 1944
Remarks: The reactivation of possibly four 'Iowa' class battleships marks an extraordinary renaissance for the type, whose capabilities are needed as an interim measure pending the delivery of more aircraft-carriers to the US Navy. The initial modernization upgrades the electronics fit, provides medium and long-range anti-ship capability with Harpoon and Tomahawk missiles, improves point defence with four Phalanx CIWS mountings, and adds an anti-submarine capability in the form of helicopters.

CALIFORNIA CLASS
NUCLEAR-POWERED GUIDED-MISSILE CRUISER

Country of Origin: USA

Displacement: 9,560 tons standard and 11,100 tons full load

Dimensions: Length 181.7m (596ft); beam 18.6m (61ft); draught 9.6m (31.5ft)

Gun Armament: Two 127mm (5in) L/54 DP in two Mk 42 single mountings, and (to be fitted) two Phalanx Mk 15 20mm (0.8in) close-in weapons system mountings

Missile Armament: Two quadruple container-launchers for eight BGM-109 Tomahawk surface-to-surface missiles, two quadruple launchers for eight RGM-84A Harpoon surface-to-surface missiles, and two Mk 13 single launchers for 80 RIM-66C Standard MR surface-to-air missiles

Anti-submarine Armament: One octuple Mk 16 launcher for RUR-5A ASROC missiles, and two triple Mk 32 tube mountings for 324mm (12.75in) Mk 46 A/S torpedoes

Propulsion: Two General Electric D2G pressurized-water cooled reactors supplying steam to two geared

turbines delivering 44,740kW (60,000shp) to two shafts
Performance: Maximum speed 30kts; nuclear core life about 1,126,500km (700,000mls)
Complement: 28+512
Class
1. USA (2)
CALIFORNIA Commissioned February 1974
SOUTH CAROLINA Commissioned January 1975
Remarks: Designed to constitute the primary area-defence escort of a nuclear-powered aircraft carrier, the ships proved very expensive to build, and this led to the cancellation of a proposed third unit. In US Navy terminology these are 'double-ended' ships, with missile armament at each end of the flush deck, separated by the superstructure and enclosed tower masts.

KIROV CLASS
NUCLEAR-POWERED BATTLE CRUISER

Country of Origin: USSR

Displacement: 22,000 tons standard and 28,000 tons full load

Dimensions: Length 250.0m (820.2ft); beam 28.5m (93.5ft); draught 10.0m (32.8ft)

Gun Armament: Two 100mm (3.9in) DP in single mountings, and eight 30mm (1.2in) AA 'Gatling' mountings

Missile Armament: 20 launch tubes for 20 SS-N-19 surface-to-surface missiles, 12 launch tubes for 96 SA-N-6 surface-to-air missiles, and two twin launchers for 36 SA-N-4 'Gecko' surface-to-air missiles

Torpedo Armament: Two quintuple 533mm (21in) tube mountings

Anti-submarine Armament: One RBU 6000 12-barrel rocket-launcher, two RBU 1000 six-barrel rocket launchers, one twin launcher for 16 SS-N-14 'Silex' anti-submarine missiles, and helicopter-launched weapons

Aircraft: Three to five Kamov KA-25 'Hormone' helicopters on a platform aft
Electronics: One 'Top Pair' 3D radar, one 'Top Steer' 3D radars, two 'Eye Bowl' SS-N-14 control radars, two

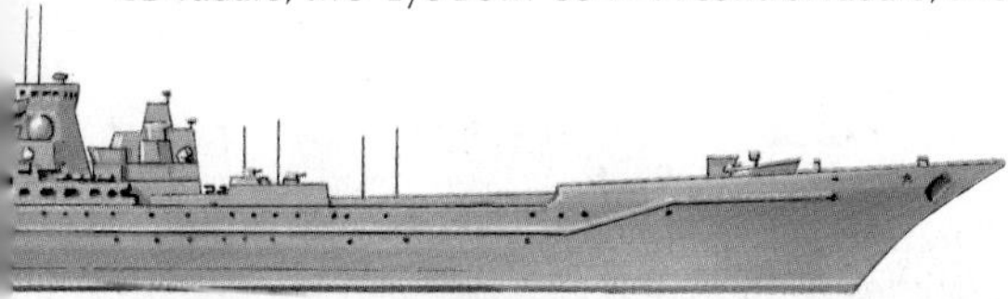

'Top Dome' SA-N-6 control radars, two 'Pop Group' SA-N-4 control radars, one 'Kite Screech' main armament control radars, four 'Bass Tilt' AA gun-control radars, three 'Palm Frond' navigation radars, hull-mounted sonar, variable-depth sonar, and a comprehensive electronic countermeasure suite including eight 'Side Globe', four 'Rum Tub' and two 'Round House' antenna/housings
Propulsion: Two nuclear reactors with boilers supplying steam to turbines delivering 119,300kW (160,000shp)
Performance: Maximum speed 35kts
Complement: 900
Class
1. USSR (1+1)
KIROV Commissioned..1981
Remarks: One of the most remarkable classes to have appeared in recent years, the Kirov class battle cruiser is designed for independent operations.

KRASINA CLASS GUIDED-MISSILE CRUISER

Country of Origin: USSR

Displacement: 11,000 tons standard and 13,000 tons full load

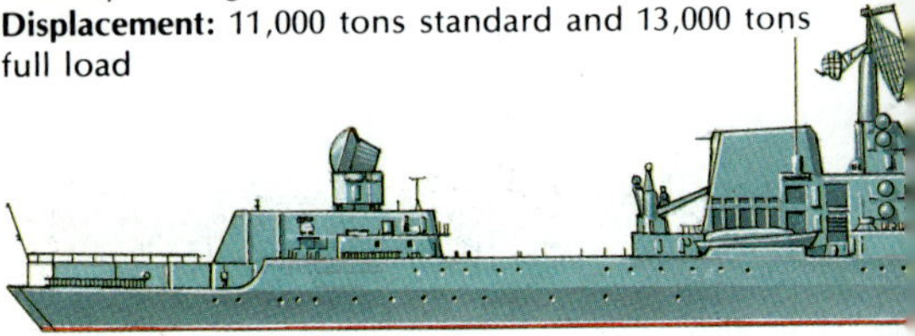

Dimensions: Length 187.0m (613.5ft); beam 22.3m (73.2ft); draught 7.6m (25ft)

Gun Armament: Two 130mm (5.12in) L/70 in one twin mounting, and six 30mm (1.2in) 'Gatling' type CIWS mountings

Missile Armament: Eight twin container-launchers for 16 SS-N-12 surface-to-surface missiles, and two twin launch tubes for 84 SA-N-6 surface-to-air missiles, and two twin launchers for 40 SA-N-4 surface-to-air missiles

Torpedo Armament: Two quadruple or quintuple tube mountings for dual-role anti-submarine torpedoes

Anti-submarine Armament: Two 12-barrel RBU 6000 launchers, and anti-submarine torpedoes (above)

Aircraft: One Kamov KA-25 'Hormone-B' over-the-horizon targeting helicopter in a hangar aft

Electronics: One 'Top Pair' 3D radar, one 'Top Steer' 3D radar, three 'Palm Frond' navigation and surface-

search radars, one 'Trap Door' type SSM-control radar, one 'Top Dome' SA-N-6 control radar, two 'Pop Group' SA-N-4 control radars, one 'Kite Screech' 130mm (5.12in) fire- control radar, three 'Bass Tilt' 30mm fire-control radars, two 'Tee Plinth' IR surveillance systems, eight 'Side Globe' ECM fairings, four 'Rum Tub' ESM fairings, ? 'Bell' series ESM fairings, ? 'High Pole-B' IFF, two chaff launchers, and one low- frequency hull sonar, and one medium-frequency variable-depth sonar
Propulsion: COGOG (COmbined Gas turbine Or Gas turbine) arrangement with four gas turbines delivering about 103,000kW (174,330shp) to two shafts
Performance: Maximum speed 34kts
Complement: About 600
Class
1. USSR (2+1)
SLAVA Commissioned 1983

LONG BEACH
NUCLEAR-POWERED GUIDED-MISSILE CRUISER

Country of Origin: USA

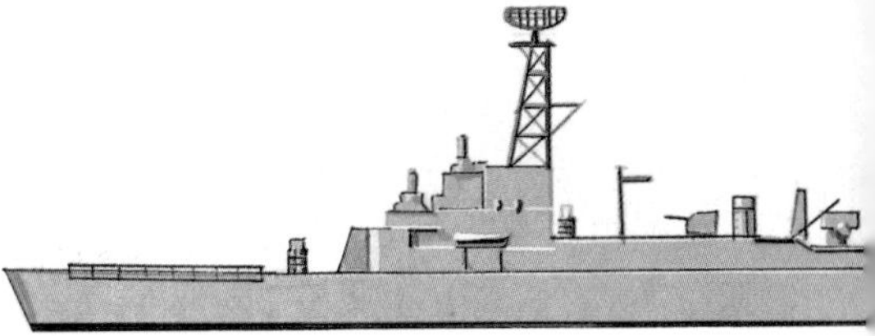

Displacement: 14,200 tons standard and 17,100 tons full load

Dimensions: Length 219.9m (721.1ft); beam 22.3m (73.2ft); draught 9.1m (26.7ft)

Gun Armament: Two 127mm (5in) L/38 DP in Mk 30 single mountings, and two 20mm (0.8in) Phalanx Mk 15 close-in weapon system mountings

Missile Armament: Two quadruple container-launchers for eight RGM-84A anti-ship missiles, two Mk 10 twin launchers for RIM-67B Standard-ER surface-to-air missiles

Anti-submarine Armament: Two triple Mk 32 tube mountings for 324mm (12.75in) Mk 46 A/S torpedoes, and one octuple launcher for RUR-5A ASROC missiles

Aircraft: Provision for a helicopter on a platform aft

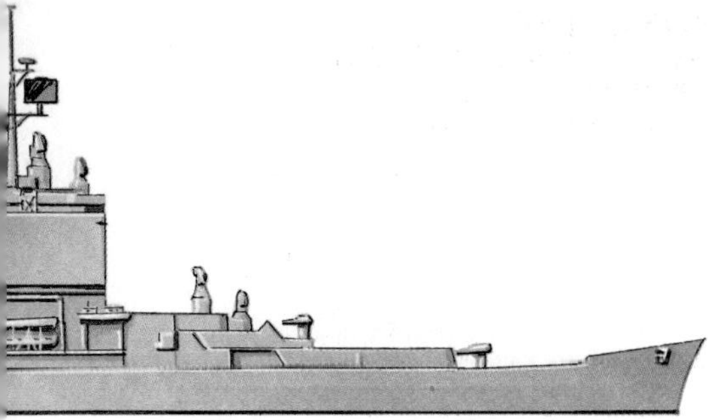

sets of General Electric geared turbines delivering 59,655kW (80,000shp) to two shafts
Performance: Maximum speed 36kts
Complement: 79+1,081, plus flag accommodation for 10+58
Class
1. USA (1)
LONG BEACH Commissioned September 1961
Remarks: The *Long Beach* was the world's first nuclear-powered surface combatant, and the first ship designed for the US Navy specifically as a cruiser since World War II. She was also the world's first warship to have a missile main armament. Recently modernized and refurbished, the *Long Beach* has the performance, size and communications to operate effectively as a flagship.

MOSKVA CLASS
HELICOPTER CRUISER

Country of Origin: USSR
Displacement: 16,500 tons standard and 20,000 tons full load
Dimensions: Length 190.5m (624.8ft); beam 23.0m (75.4ft); draught 11.0m (36.1ft); flight-deck width 34.0m (111.5ft)

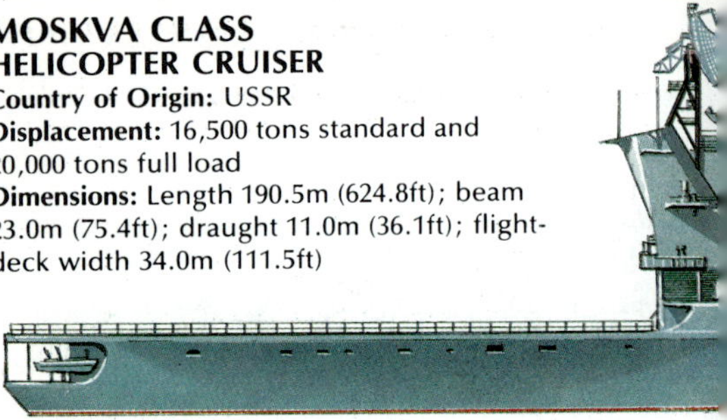

Gun Armament: Four 57mm L/70 AA in two twin mountings
Missile Armament: Two twin launchers for 48 SA-N-3; 'Goblet' surface-to-air missiles
Anti-submarine Armament: One twin SUW-N-1 launcher for 20 FRAS-1 missiles, two RBU 6000 12-barrel rocket-launchers, and helicopter-launched weapons
Aircraft: 18 Kamov Ka-25 'Hormone-A' helicopters
Electronics: One 'Top Sail' 3D radar, one 'Head Net-C' 3D radar, two 'Head Light' SAM-control radars, two 'Muff Cob' AA gun-control radars, three 'Don-2' navigation radars, one low-frequency hull-mounted sonar, one medium-frequency variable-depth sonar, and an extensive electronics countermeasures suite including eight 'Side Globe' antennae/housings
Propulsion: Four boilers supplying steam to two sets

of geared turbines delivering 74,570kW (100,000shp) to two shafts
Performance: Maximum speed 30kts; range 16,675km (10,360mls) at 18kts or 5,200km (3,230mls) at 30kts

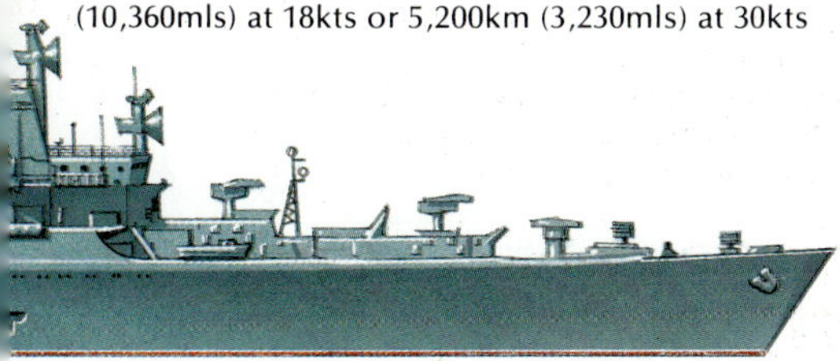

Complement: 840 excluding air group personnel
Class
1. USSR (2)
MOSKVA Commissioned May 1967
LENINGRAD Commissioned late 1966
Remarks: The Moskva class marked a big step forward in Soviet maritime air power and anti-submarine capabilities, but was nonetheless something of a mixed blessing. For while anti-submarine and task-group command problems were eased quite considerably by the availability of these two ships, their seakeeping qualities leave much to be desired, and it was also realized as the ships were being built that Soviet warships needed much-enhanced anti-ship capability. Thus only two of a planned 20 units were built, the class being replaced by the more capable Kiev class aircraft carriers.

ANDREA DORIA CLASS
GUIDED-MISSILE CRUISER

Country of Origin: Italy

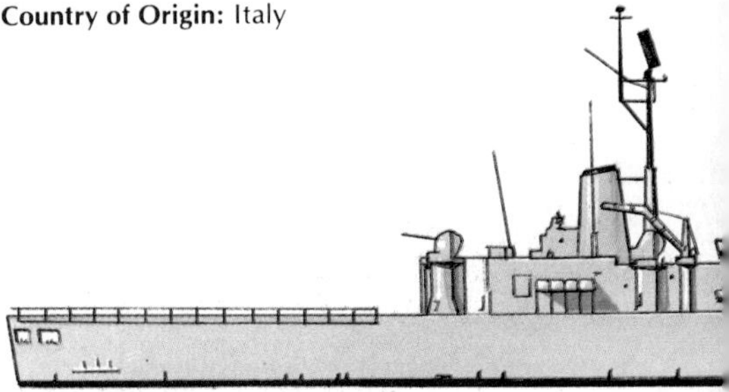

Displacement: 5,000 tons standard and 6,500 tons full load
Dimensions: Length 149.3m (489.8ft); beam 17.2m (56.4ft); draught 5.0m (16.4ft)
Gun Armament: Eight 76mm (3in) OTO-Melara L/62 in single mountings
Missile Armament: One twin launcher for 40 RIM-67A Standard-ER surface-to-air missiles
Anti-submarine Armament: Two triple MK 32 tube mountings for 324mm (12.75in) MK 46 A/S torpedoes, and helicopter-launched torpedoes
Aircraft: Four Agusta-Bell AB.212 ASW helicopters
Propulsion: Four Foster-Wheeler boilers supplying steam to two double-reduction geared turbines delivering 44.740kW (60,000shp) to two shafts

Performance: Maximum speed 31kts; range 9,250km (5,750mls) at 17kts

Complement: 45+425
Class
1. Italy (2)
ANDREA DORIA Commissioned February 1964
CAIO DUILIO Commissioned November 1964
Remarks: The Andrea Doria class represents Italian innovative capability at its best, combining many of the attributes of the guided-missile destroyer and helicopter cruiser on a hull of moderate displacement. The squeeze was a little too tight, and conditions for the Agusta-Bell AB 212ASWs are cramped to the extent that operational efficiency suffers. The helicopter platform measures 30.0m by 16.0m (98.5ft by 52.5ft), with a hangar in the after superstructure just forward of the helicopter platform.

BAINBRIDGE CLASS
NUCLEAR-POWERED GUIDED-MISSILE CRUISER

Country of Origin: USA

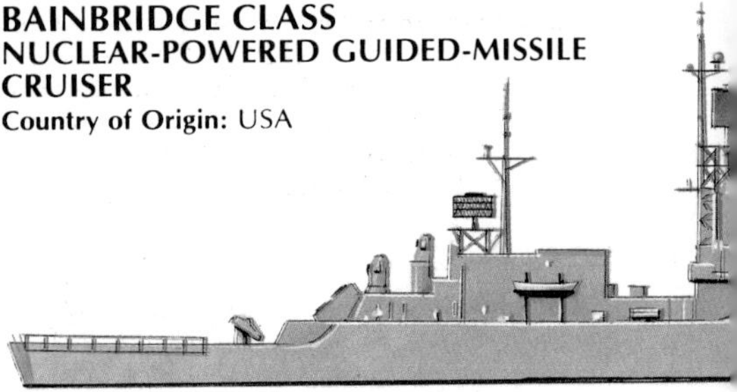

Displacement: 7,600 tons standard and 8,592 tons full load

Dimensions: Length 172.3m (565ft); beam 17.6m (57.9ft); draught 7.7m (25.4ft)

Gun Armament: Two 20mm (0.8in) AA in a MK 67 twin mounting

Missile Armament: Two quadruple launchers for RGM-84A Harpoon surface-to-surface missiles, and two MK 10 twin launchers for 80 RIM-67B Standard ER surface-to-air missiles

Anti-submarine Armament: One octuple RUR-5A ASROC launcher and two triple MK 32 tube mountings for 324mm (12.75in) MK 44/46 torpedoes

Electronics: One Hughes SPS-52 3D air-search radar, one Raytheon/Sylvania SPS-10 surface-search radar, one Raytheon SPS-49 air-search radar, four Sperry/RCA SPG-55A fire control radars used in conjunction with four MK 76 missile fire-control systems, one MK

11 weapons-direction system (to be replaced by 14 weapons-direction system), URN-20 TACAN, Naval Tactical Data System, OE-82 satellite communications antenna, SRR-1 satellite transceiver, one Sangamo SQ-23 long-range active sonar, four MK 36 Super Chaffroc rapid-blooming overhead chaff system, and

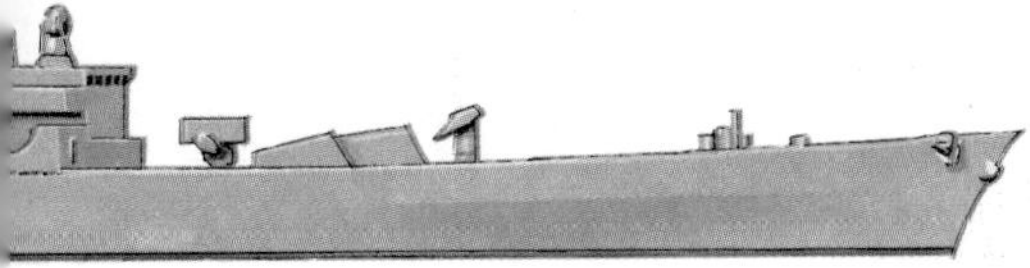

one SLQ-32(V)3 ESM suite

Propulsion: Two General Electric D2G pressurized-water cooled reactors supplying steam to two geared turbines delivering 44,740kW (60,000shp) to two shafts

Performance: Maximum speed 38kts

Complement: 34+436, plus flag accommodation for 6+12

Class

1. USA (1)

BAINBRIDGE Commissioned October 1962

Remarks: *Bainbridge* was the third of the US Navy's nuclear powered surface ships and was originally classified as a guided missile frigate before being reclassified on 30 June, 1975. whilst undergoing modernization to its current specification at the Paget Sound Naval Shipyard from June 1974 to September 1976.

BELKNAP CLASS
GUIDED-MISSILE LIGHT CRUISER

Country or Origin: USA

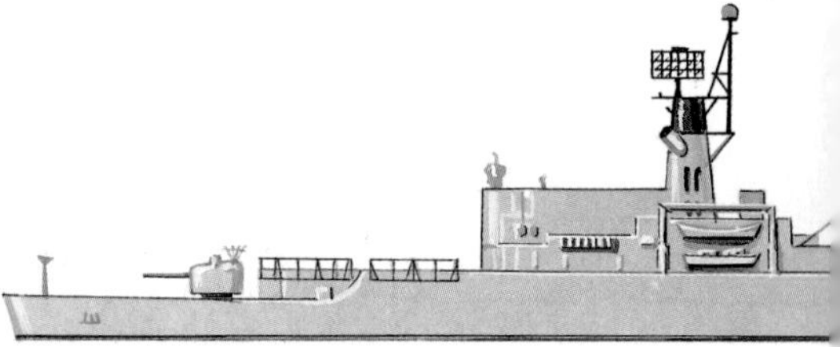

Displacement: 6,570 tons standard and 7,900 tons full load

Dimensions: Length 166.7m (547ft); beam 16.7m (54.8ft); draught 8.8m (28.8ft) to bottom of sonar dome and 5.8m (19ft) to keel

Gun Armament: One 127mm (5in) L/54 in a MK 42 mounting, and two 20mm (0.8in) Phalanx MK 15 close-in weapon system mountings

Missile Armament: Two quadruple launchers for eight RGM-84A Harpoon surface-to-surface missiles, one MK 10 twin launcher for 40 RIM-67B Standard SM-2 surface-to-air missiles, and (to be fitted) BGM-109 Tomahawk cruise missiles

Anti-submarine Armament: Two triple MK 32 tube mountings for 324mm (12.75in) MK 44/46 A/S torpedoes, and up to 20 RUR-5A ASROC missiles fired from the same MK 10 launcher as the Standard SAMs

Propulsion: Two pressurized-water cooled Westinghouse C1W nuclear reactors supplying steam to two

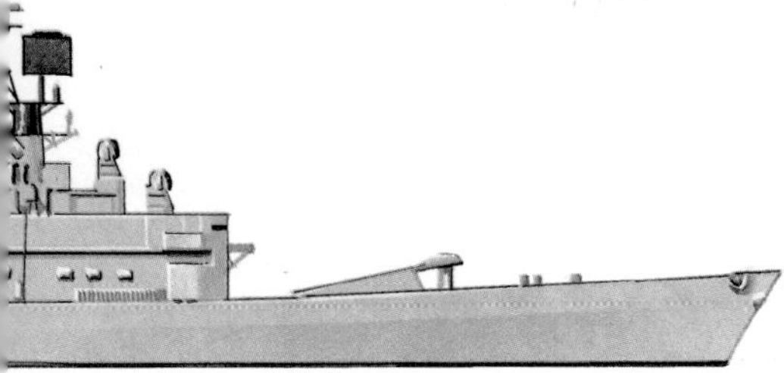

Aircraft: One Kaman SH-2F Seasprite helicopter
Propulsion: Four Babcock & Wilcox boilers supplying steam to two sets of geared General Electric turbines delivering 63,385kW (85,000shp) to two shafts
Performance: Maximum speed 32.5kts
Complement: 31+387 (including squadron staff) or 520 (CG 26 including flag accommodation)
Class 1. USA (9)
BELKNAP Commissioned November 1964
Remarks: The Belknap class of guided-missile cruisers serve with the Atlantic and Pacific Fleets (three and six units respectively), and with the Leahy class form the basis of US carrier task forces' escort capability. However, compared with the Leahy class, the Belknap class is a 'single-ended' design, and can thus engage only two aerial targets simultaneously.

KARA CLASS GUIDED-MISSILE CRUISER

Country of Origin: USSR

Displacement: 8,200 tons standard and 10,000 tons full load

Dimensions: Length 173.2mm (568ft); beam 18.0m (59ft); draught 6.7m (22ft)

Gun Armament: Four 76mm (3in) L/60 DP in two twin mountings, and four 30mm (1.2in) AA 'Gatling' mountings

Missile Armament: Two quadruple container-launchers for eight SS-N-14 surface-to-underwater missiles, two twin launchers for 72 SA-N-3 'Goblet' surface-to-air missiles, and two twin launchers for 36 SA-N-4 'Gecko' SAMs

Torpedo Armament: Two quintuple 533mm (21in) AS/ASW tube mountings

Anti-submarine Armament: Two RBU 6,000 12-barrel rocket-launchers, two RBU 1,000 six barrel rocket-launchers, and helicopter-launched weapons

Aircraft: One Kamov KA-25 'Hormone' helicopter

Propulsion: COGOG (COmbined Gas Or Gas turbine) arrangement, with two gas turbines delivering

17,900kW (24,000shp) and four gas turbines delivering 74,570kW (100,000shp) to two shafts
Performance: Maximum speed 33kts; range 14,825km (9,210mls) at 18kts or 3,700km (2,300mls) at 32kts
Complement: 30+510
Class
1. USSR (7)
NIKOLAYEV Commissioned 1971
OCHAKOV Commissioned 1973
KERCH Commissioned 1974
AZOV Commissioned 1975
PETROPAVLOVSK Commissioned 1976
TASHKENT Commissioned 1977
TALLINN Commissioned 1979
Remarks: The 'Kara' class was the first type of Soviet heavy cruiser built since the 'Sverdlov' class, and the seven ships of the class represent a formidable anti-submarine capability with long range and a mix of medium and short-range surface-to-air missiles for area and point defence.

KRESTA I CLASS
GUIDED-MISSILE CRUISER
Country of Origin: USSR

Displacement: 6,140 tons standard and 7,500 tons full load
Dimensions: Length 155.5m (510ft); beam 17.0m (55.7ft); draught 6.0m (19.7ft)
Gun Armament: Four 57mm (2.25in) L/80 DP in two twin mountings
Missile Armament: Two twin container-launchers for four SS-N-3B 'Shaddock' surface-to-surface missiles, and two twin launchers for 32 SA-N-1 'Goa' surface-to-air missiles
Torpedo Armament: Two quintuple 533mm (21in) AS/ASW tube mountings
Anti-submarine Armament: Two RBU 6,000 12-barrel rocket-launchers, and two RBU 1,000 six- barrel rocket-launchers
Aircraft: One Kamov Ka-25 'Hormone-B' helicopter in a hangar aft
Electronics: One 'Head Net-C' 3D radar, one 'Big Net' air-search radar, two 'Plinth Net' surface-search

radars, one 'Scoop Pair' SSM-control radar, two 'Peel Group' SAM-control radars, two 'Muff Cob' main armament control radars, two 'Don-2' navigation

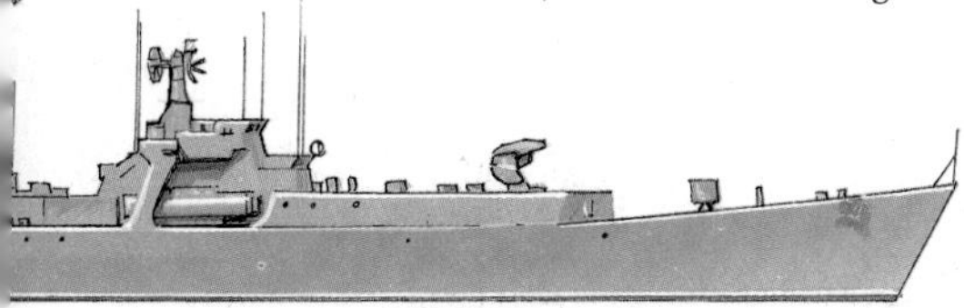

radars, one 'High Pole-B' IFF, one hull-mounted sonar, and a comprehensive electronic counter-measures system with eight 'Side Globe' antennae/housings

Propulsion: Four boilers supplying steam to two sets of geared turbines delivering 74,570kW (100,000shp) to two shafts

Performance: Maximum speed 34kts; range 13,000km (8,080mls) at 18kts or 3,700km (2,300mls) at 32kts

Complement: 375

Class

1. USSR (4)

ADMIRAL ZOZULYA Commissioned 1967

Remarks: The Kresta I class was designed to supplant the Kynda class for anti-ship warfare, and was the first Soviet cruiser class designed with provision for an embarked over-the-horizon targeting helicopter, giving the units of the class the ability to operate effectively beyond the range of Soviet land-based aircraft.

KYNDA CLASS
GUIDED-MISSILE CRUISER

Country of Origin: USSR

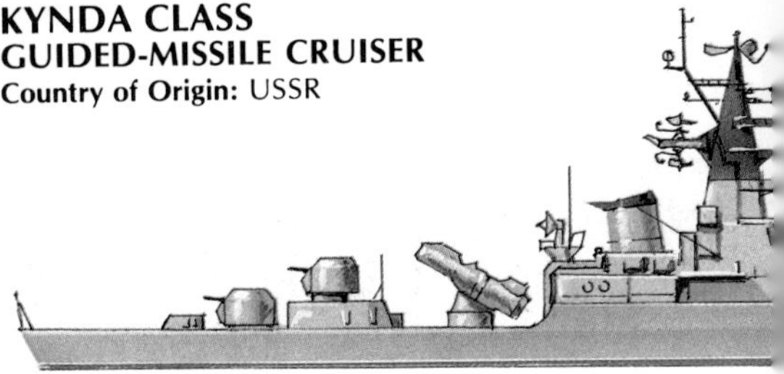

Displacement: 4,400 tons standard and 5,700 tons full load
Dimensions: Length 142.0m (465.8ft); beam 15.8m (51.8ft); draught 5.3m (17.4ft)
Gun Armament: Four 76mm (3in) L/60 DP in two mountings
Missile Armament: Two quadruple container-launchers for 16 SS-N-3B 'Shaddock' surface-to-surface missiles, and one twin launcher for 22 SA-N-1 'Goa' surface-to-surface missiles
Torpedo Armament: Two triple 533mm (21in) AS/ASW tube mountings
Anti-submarine Armament: Two RBU 6000 12-barrel rocket-launchers
Aircraft: Provision for one Kamov Ka-25 'Hormone' helicopter on a platform aft
Electronics: Two 'Head Net-A' air-search radars, two 'Plinth Net' surface-search radars, two 'Scoop Pair'

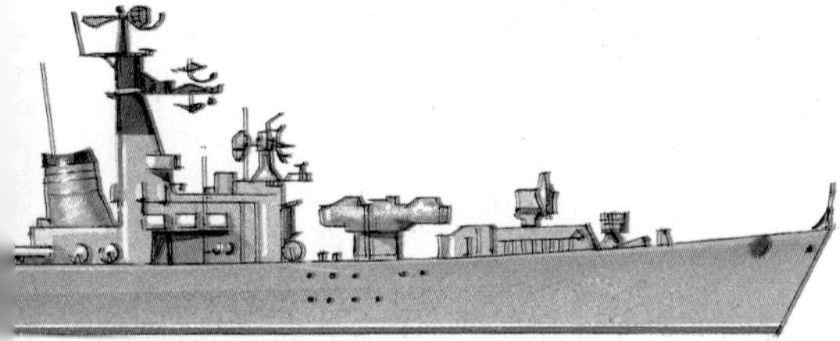

SSM-control radars, one 'Peel Group' SAM-control radar, two 'Owl Screech' gun-control radars, two 'Don-2' navigation radars, one 'High Pole-B' IFF, one hull-mounted sonar, and various electronic warfare systems

Propulsion: Four boilers supplying steam to two sets of geared turbines delivering 74,570kW (100,000shp) to two shafts

Performance: Maximum speed 34kts; range 11,125km (6,915mls) at 15kts or 2,775km (1,725mls) at 34kts

Complement: 390

Class

1. USSR (4)

Remarks: The Kynda class was the USSR's first guided-missile cruiser class, and its units were optimized for anti-ship warfare in the role of destroyer flotilla leader. The missile armament is still useful, but reloading of the SS-N-3 tubes is slow.

LEAHY CLASS GUIDED-MISSILE CRUISER

Country of Origin: USA

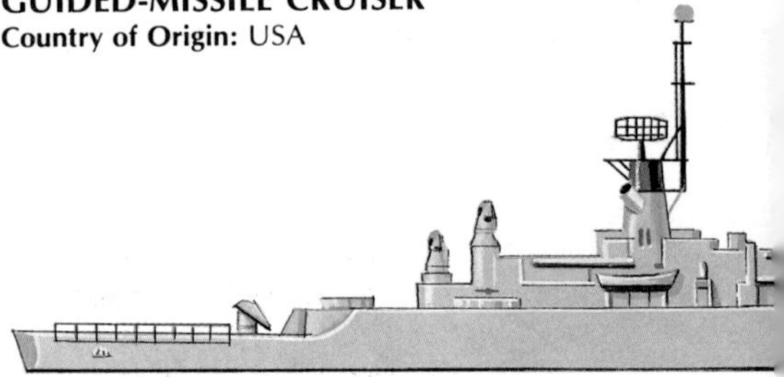

Displacement: 5,670 tons standard and 7,800 tons full load
Dimensions: Length 162.5m (533ft); beam 16.6m (54.9ft); draught 7.6m (24.8ft) to sonar dome
Gun Armament: Two 20mm (0.79in) Phalanx Mk 15 close-in weapon system mountings
Missile Armament: Two quadruple container-launchers for eight RGM-84A Harpoon surface-to-surface missiles, and two Mk 10 twin launchers for 80 RIM-67B Standard-ER surface-to-air missiles
Anti-submarine Armament: Two triple Mk 32 tube mountings for 324mm (12.75in) Mk 46 A/S torpedoes, and one octuple launcher for RUR-5A ASROC missiles
Aircraft: Provision for one Kaman SH-2F Seasprite on a platform aft
Propulsion: Four Babcock & Wilcox boilers supplying

steam to two sets of General Electric geared turbines delivering 63,385kW (85,000shp) to two shafts
Performance: Maximum speed 32.7kts; range

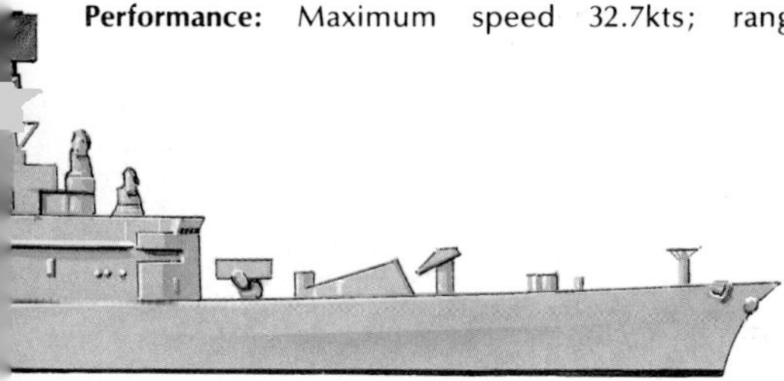

14,805km (9,200mls) at 20kts
Complement: 18+359 plus flag accommodation for 6+18
Class
1. USA (9)
LEAHY Commissioned August 1962
HARRY E YARNELL Commissioned February 1963
WORDEN Commissioned August 1963
DALE Commissioned November 1963
RICHMOND K TURNER Commissioned June 1964
GRIDLEY Commissioned May 1963
ENGLAND Commissioned December 1963
HALSEY Commissioned July 1963
REEVES Commissioned May 1964
Remarks: The Leahy class was the US Navy's first purpose-built guided-missile cruiser class, and was designed for the escort of fast carrier task groups.

SOVREMENNY CLASS
GUIDED-MISSILE CRUISER

Country of Origin: USSR

Displacement: 6,200 tons standard and 7,800 tons full load
Dimensions: Length 156.0m (511.8ft); beam 17.3m (56.8ft); draught 6.5m (21.3ft)
Gun Armament: Four 130mm (5.1in) L/60 DP in two twin mountings, and four 30mm (1.18in) AA in 'Gatling' mountings
Missile Armament: Two quadruple container-launchers for eight SS-N-22 anti-ship missiles, and two launchers for 48 SA-N-7 surface-to-air missiles
Torpedo Armament: Two twin 533mm (21in) AS/ASW tube mountings
Anti-Submarine Armament: Two RBU 1000 six-barrel rocket-launchers
Aircraft: One Kamov Ka-25 'Hormone' helicopter in a hangar amidships
Electronics: One 'Top Steer' 3D radar, one 'Band Stand' SSM-control radar, six 'Front Dome' missile-control radars, one 'Kite Screech' main armament

gun-control radar, two 'Bass Tilt' AA gun-control radars, three 'Palm Frond' navigation radars, extensive electronic countermeasures systems, and hull-

mounted sonar
Propulsion: ? boilers supplying steam to two sets of turbo-pressurized turbines delivering 74,570kW (100,000shp) to two shafts
Performance: Maximum speed 34kts
Complement: About 350
Class
1. USSR (4+2)
SOVREMENNY Commissioned August 1980
OTCHYANNY Commissioned 1982
Remarks: Originally classified by NATO as BAL COM2, these craft were purpose-built for surface warfare and are complementary to the Udaloy ASW Class. Built in Leningrad by the Zhdanor Yard, engine tests were carried out in the Baltic in 1980 following which the missiles and guns were fitted. The helicopter is carried on a telescopic hangar. *Sovremenny* was transferred to the Northern Fleet in early 1982 where she was joined by *Otchyanny* later in the year.

TICONDEROGA CLASS GUIDED-MISSILE CRUISER

Country of Origin: USA
Displacement: 9,600 tons full load
Dimensions: Length 172.8m (566.8ft); beam 16.8m (55ft); draught 9.5m (31ft)

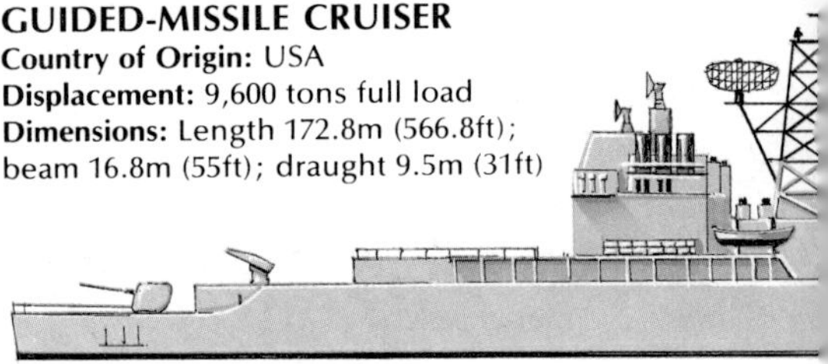

Gun Armament: Two 127mm (5in) L/54 DP in two Mk 45 single mountings, and two 20mm (0.79in) Phalanx Mk15 close-in weapon system mountings
Missile Armament: Two octuple container-launchers for 16 RGM-84A Harpoon surface-to-surface missiles, and two Mk 26 twin launchers for up to 68 RIM-67B Standard surface-to-air missiles; from CG49 onwards the ships will have two EX 41 vertical launchers in place of the Mk 26 launchers for up to 122 assorted missiles, and from CG 52 onwards the ships will have two Vertical Launch Systems each with 12 BGM-109 Tomahawk surface-to-surface cruise missiles
Anti-submarine Armament: Two triple Mk 32 tube mountings for 324mm (12.75in) Mk 46 A/S torpedoes, up to 20 RUR-5A ASROC missiles forming part of the total missile strength above and launched from the Mk 26 launchers, and helicopter-launched weapons
Aircraft: Two Sikorsky SH-60B Seahawk helicopters

Propulsion: Four General Electric LM 2500 gas tur-

bines delivering 59,655kW (80,000shp) to two shafts
Performance: Maximum speed 30+kts
Complement: 33+327
Class
1. USA (2+9+14)
TICONDEROGA Commissioned January 1983
YORKTOWN Commissioned July 1984
Remarks: The Ticonderoga class guided-missile cruiser is one of the most significant classes to have appeared in US Navy service in recent years, and is based on a slightly stretched version of the hull and powerplant of the Spruance class guided-missile destroyer. The key to the ships' importance is the AEGIS system fitted in each vessel: this combines the SPY-1A long-range radar (with four mechanically-fixed but electronically-scanning antennae) for hemispherical detection, acquisition and tracking of multiple targets, with the AEGIS Weapons Control System Mk 7 with its UYK-7 computers.

TRUXTUN CLASS
NUCLEAR-POWERED GUIDED-MISSILE CRUISER

Country of Origin: USA

Displacement: 8,200 tons standard and 9,125 tons full load

Dimensions: Length 171.9m (564ft); beam 17.7m (58ft); draught 9.4m (31ft)

Gun Armament: One 127mm (5in) L/54 DP in one Mk 24 single mounting, and two 20mm (0.79in) Phalanx Mk 15 close-in weapon system mountings

Missile Armament: Two quadruple container-launchers for eight RGM-84A Harpoon surface-to-surface missiles, and one Mk 10 twin launcher for up to 60 RIM-67B Standard-ER surface-to-air missiles

Anti-submarine Armament: Four Mk 32 tubes for 324mm (12.75in) Mk 46 A/S torpedoes, up to 20 RUR-5A ASROC missiles included in the total above for the Mk 10 launcher and launched from the Mk 10 launcher, and helicopter-launched weapons

Aircraft: One Kaman SH-2F Seasprite helicopter

Electronics: One ITT/Gilfillan SPS-48 3D radar, one Lockheed SPS-40 air-search radar, one Raytheon/Sylvania SPS-10 surface-search radar, one SPG-53F gun-control radar used in conjunction with the Mk 68 gun fire-control system, two Sperry/RCA SPG-55B SAM-control radars used in conjunction with the two Mk 76 missile fire-control systems, one Mk 14 weapon

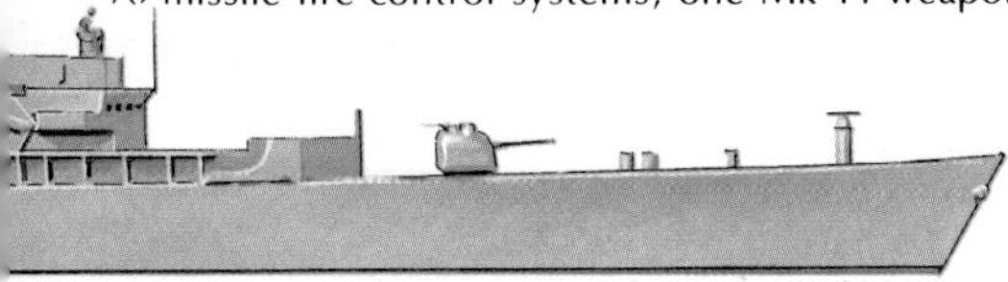

direction system, one EDO/General Electric SQS-26 bow-mounted 'bottom-bounce' sonar
Propulsion: Two pressurized-water cooled General Electric D2G nuclear reactors supplying steam to two sets of geared turbines delivering 44,740kW (60,000shp) to two shafts
Performance: Maximum speed 38kts
Complement: 36+492, plus flag accommodation for 6+12
Class
1. USA (1)
TRUXTUN Commissioned May 1967
Remarks: *Truxton* was the US Navy's fourth nuclear-powered surface warship and was based on the Belknap oil-burning frigate design. Originally classified as a guided missile frigate it was reclassified as a guided-missile cruiser on 30 June 1975.

TYPE 82 CLASS
GUIDED-MISSILE CRUISER
Country of Origin: UK

Displacement: 6,100 tons standard and 7,100 tons full load
Dimensions: Length 154.5m (507ft); beam 16.8m (55ft); draught 7.0m (23ft) to sonar bome
Gun Armament: One 114mm (4.5in) L/55 DP in one Mk8 single mounting, and two 20mm (0.79in) Oerlikon AA in two Mk 7 single mountings
Missile Armament: One twin launcher for 40 Sea Dart surface-to-air missiles
Anti-Submarine Armament: One launcher for 40 Ikara missiles, one Limbo Mk 10 mortar, and helicopter-launched weapons
Aircraft: One Westland Wasp HAS.Mk 1 helicopter on a platform aft
Propulsion: COSAG (COmbined Steam And Gas turbine) arrangement, with two boilers supplying steam to two sets of Admiralty Standard Range turbines

delivering 22,370kW (30,000shp) and two Rolls-Royce Olympus TM1A gas turbines delivering 22,370kW ((30,000shp) to two shafts
Performance: Maximum speed 28kts; range 9,255km (5,750mls) at 18kts
Complement: 29+378
Class
1. UK (1)
Name:
BRISTOL Commissioned March 1973
Remarks: Following the cancellation of the aircraft-carrier building programe, due to high costs, the Type 82 class which were intended as escort vessels were also cancelled. Classified as a destroyer due to her Sea Dart area defence capability, although the Type 82 does not carry helicopters. She was refitted in 1979-80.

UDALOY CLASS
GUIDED-MISSILE CRUISER

Country of Origin: USSR

Displacement: 6,700 tons standard and 8,200 tons full load

Dimensions: Length 162.0m (531.5ft); beam 19.3m (63.3ft); draught 6.2m (20.3ft)

Gun Armament: Two 100mm (3.9in) DP in single mountings, and four 30mm (1.18in) AA in 'Gatling' mountings

Missile Armament: Two quadruple container-launchers for eight SS-N-14 surface-to-underwater and surface-to-surface missiles, and eight launchers for 48 SA-N-8 surface-to-air missiles

Torpedo Armament: Two quadruple 533mm (21in) AS/ASW tube mountings

Anti-submarine Armament: Two RBU 6000 12-barrel rocket-launchers, SS-N-14 missiles, and helicopter-launched weapons

Aircraft: Two Kamov Ka-27 'Helix-A' helicopters in two hangars aft

Electronics: Two 'Strut Pair' air-search radars, two 'Eye Bowl' missile-control radars, one 'Kite Screech' main armament gun-control radar, two 'Bass Tilt' AA gun-control radars, three 'Palm Frond' navigation radars, two 'High Pole-B' IFF, bow-mounted sonar, variable-depth sonar, and extensive electronic counter-measure systems
Propulsion: COGOG (COmbined Gas turbine Or Gas turbine) arrangement, with four gas turbines delivering a total of about 70,000kW (93,870shp) to two shafts
Performance: Maximum speed 35kts
Complement: About 350
Class
1. USSR (5+1)
UDALOY Commissioned 1980
VICE-ADMIRAL KULAKOV Commissioned 1981
Remarks: The Udaloy class is a potent successor to the Kresta II class anti-submarine cruiser, and is designed to work with the Sovremenny anti-ship class.

VIRGINIA CLASS
NUCLEAR-POWERED GUIDED-MISSILE CRUISER

Country of Origin: USA

Displacement: 8,625 tons standard and 10,400 tons full load
Dimensions: Length 178.4m (585ft); beam 19.2m (63ft); draught 9.0m (29.5ft)
Gun Armament: Two 127mm (5in) L/54 DP in two Mk 45 single mountings, and two 20mm (0.79in) Phalanx close-in weapon system mountings
Missile Armament: Two quadruple container-launchers for eight RGM-84A Harpoon surface-to-surface missiles, two Mk 26 twin launchers for a maximum of 50 RIM-67B Standard-MR surface-to-air missiles, and two quadruple launchers for eight BGM-109 Tomahawk cruise missiles
Anti-submarine Armament: Two triple Mk 32 tube mountings for 324mm (12.75in) Mk 46 A/S torpedoes, up to 20 RUR-5A ASROC missiles launched from the

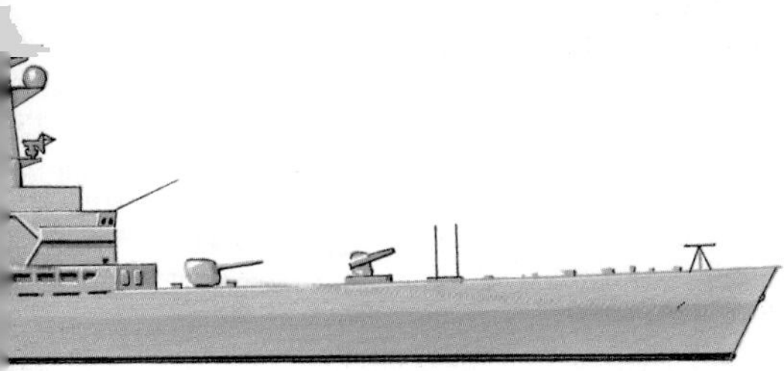

Mk 26 launchers, and helicopter-launched weapons

Aircraft: Two Sikorsky SH-60 Seahawk helicopters

Propulsion: Two pressurized-water cooled General Electric D2G nuclear reactors supplying steam to two sets of geared turbines delivering 74,570kW (100,000shp) to two shafts

Performance: Maximum speed 40kts

Complement: 27+445

Class

1. USA (4)

VIRGINIA Commissioned September 1976

Remarks: The Virginia class guided-missile cruiser is a distinct improvement in comparison with the preceding California class thanks to the provision of considerably higher speed combined with an enhanced weapon system.

ALLEN M. SUMNER CLASS DESTROYER

Country of Origin: USA/Argentina

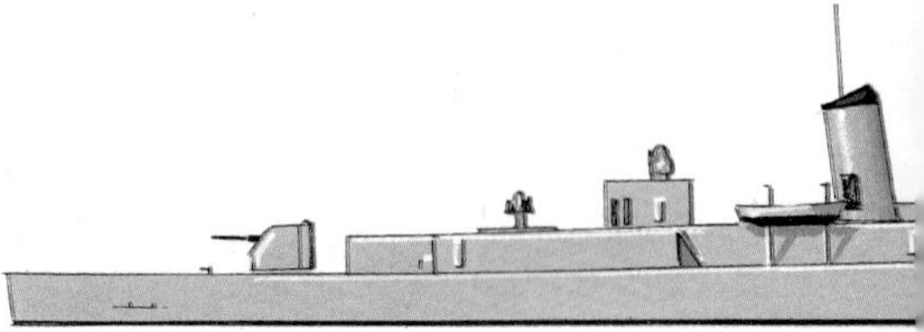

Displacement: 2,200 tons standard and 3,320 tons full load

Dimensions: Length 114.8m (376.5ft); beam 12.5m (40.9ft); draught 5.8m (19ft)

Gun Armament: Six 127mm (5in) L/38 mountings and four 76mm (3in) L/50 in two twin Mk 38 mountings

Missile Armament: Four MM38 Exocet surface-to-surface missiles in single launchers

Anti-submarine Armament: Two triple ILAS 3 tube mountings for 324mm (12.75in) Whitehead A 244S A/S torpedoes, and two forward-firing Hedgehog mortars

Propulsion: Four Babcock & Wilcox boilers supplying steam to two geared turbines delivering 44,740kW (60,000shp) to two shafts

Performance: Maximum speed 34kts; range 8,530km (5,300mls) at 15kts or 1,835km (1,140mls) at 31kts

Complement: 331

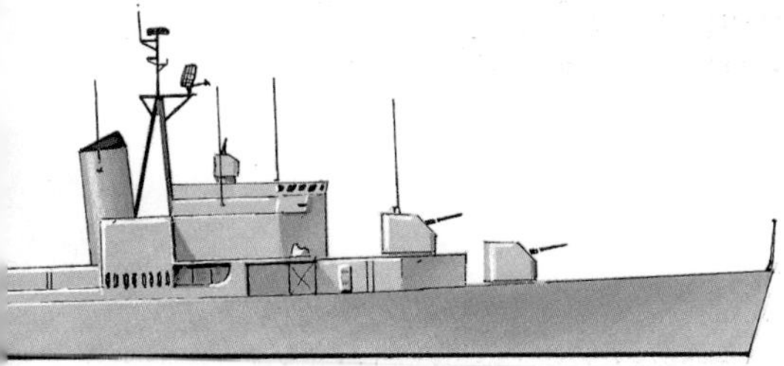

Class
1. Argentina (1)
SEGUI Commissioned August 1944
2. Brazil (1)
MATO GROSSO Commissioned November 1944
3. Taiwan (6)
PO YANG Commissioned June 1944

Remarks: The *Mato Grosso* differs from the *Segui* in having Sea Cat SAMs rather than Exocet anti-ship missiles, and Mk 32 anti-submarine torpedo tubes. The Taiwanese ships are basically similar to the Argentine vessel, but three carry Hsiung Feng (Gabriel) anti-ship missiles. None of these ships would be able to tackle a modern warship, but local conditions render them all relatively useful. The Taiwanese, in particular, face only a very modest threat from mainland China despite the acrimony of their relationship.

AUDACE CLASS
GUIDED-MISSILE DESTROYER

Country of Origin: Italy
Displacement: 3,950 tons standard and 4,560 tons full load

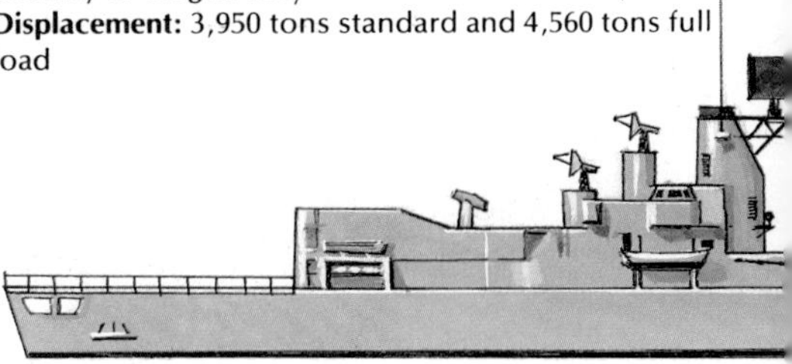

Dimensions: Length 136.6m (448ft); beam 14.2m (46.6ft); draught 4.6m (15.1ft)
Gun Armament: Two 127mm (5in) OTO-Melara Compact L/54 DP in single mountings, and four 76mm (3in) OTO-Melara Compact L/62 DP in single mountings
Missile Armament: One Mk 13 single launcher with 40 RIM-24 Tartar and RIM-66 Standard surface-to-air missile
Anti-submarine Armament: Two triple ILRS 3 tube mountings for 12 324mm (12.75in) Mk 46 A/S torpedoes, two twin 533mm (21in) tube mountings for 12 A 184 A/S torpedoes, and helicopter-launched torpedoes
Aircraft: Two Agusta-Bell AB212ASW helicopters
Electronics: One Hughes SPS-52 3D air-search radar, two Raytheon SPG-51 radars for tracking and missile guidance, one Selenia RAN 20S air-search radar, one

SMA SPQ 2 surface-search radar, three Orion RTN 10X radars used in conjunction with the ELSAG Argo NA10 gun fire-control systems, two SCLAR launcher systems for overhead flares/chaff, and one CWE 610

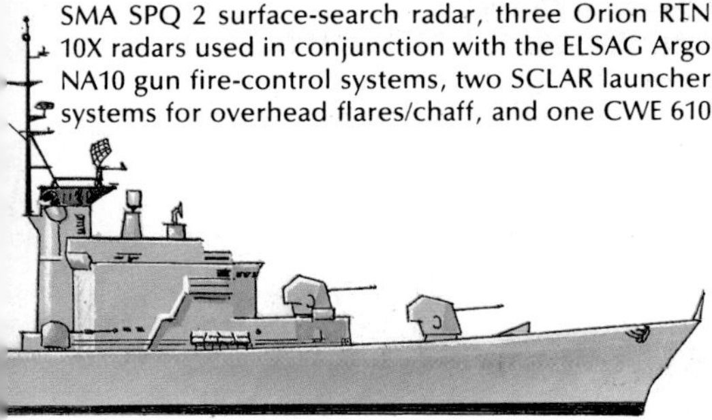

sonar

Propulsion: Four Foster-Wheeler boilers supplying steam to two double-reduction geared turbines (CNR in *Audace* and Ansaldo in *Ardito)* delivering 54,440kW (73,000shp) to two shafts

Performance: Maximum speed 34kts; range 5,560km (3,455mls) at 20kts

Complement: 30+350

Class

Italy (2+2 Audace (improved) class ships)

ARDITO Commissioned December 1973

AUDACE Commissioned November 1972

Remarks: Essentially a much improved version of the Impavido class, the Audace class provided good anti-air and anti-submarine capabilities. The two Audace (improved) class ships planned will have upgraded helicopters and anti-ship missiles.

C65 CLASS G
GUIDED-MISSILE DESTROYER

Country of Origin: France
Displacement: 3,500 tons standard and 3,900 tons full load
Dimensions: Length 127.0m (416.7ft); beam 13.4m (44ft); draught 5.8m (18.9ft)
Gun Armament: Two 100mm (3.9in) L/55 DP in single mountings
Missile Armament: Four container-launchers for single MM38 Exocet surface-to-surface missiles
Anti-submarine Armament: One launcher for 13 Malafon missiles, one quadruple 305mm (12in) mortar, and two launchers for 533mm (21in) L 5 torpedoes
Electronics: One Thomson-CSF DRBV 13 surveillance radar, one Thompson-CSF DRBV 22A air-search radar, one Thompson-CSF DRBC 32B gun fire-control radar, one Thompson-CSF DRBN 32 navigation radar,

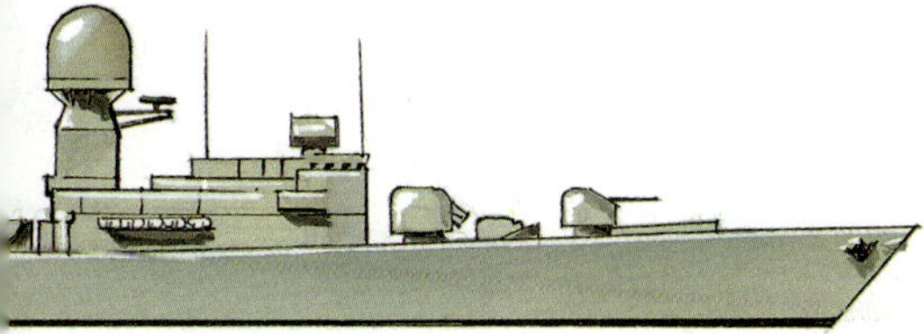

one CIT/ALCATEL DUBV 23 hull-mounted active search and attack sonar, one CTI/ALCATEL DUBV 43 variable-depth sonar, one SENIT 3 combat information system, and two Syllex eight-barrel chaff launchers

Propulsion: Two boilers supplying steam to one double-reduction Rateau geared turbine delivering 21,365kW (28,650shp) to one shaft

Performance: Maximum speed 27kts; range 9,250km (5,750mls) at 18kts

Complement: 15+89+125

Class

1. France (1)

ACONIT Commissioned March 1973

Remarks: The *Aconit* (D 609) was the prototype for the three F 67 class destroyers.

C70 CLASS
GUIDED-MISSILE DESTROYER (A/S)

Country of Origin: France
Displacement: 3,830 tons standard and 4,170 tons full load
Dimensions: Length 139.00m (455.9ft); beam 14.0m (45.9ft); draught 5.7m (18.7ft)

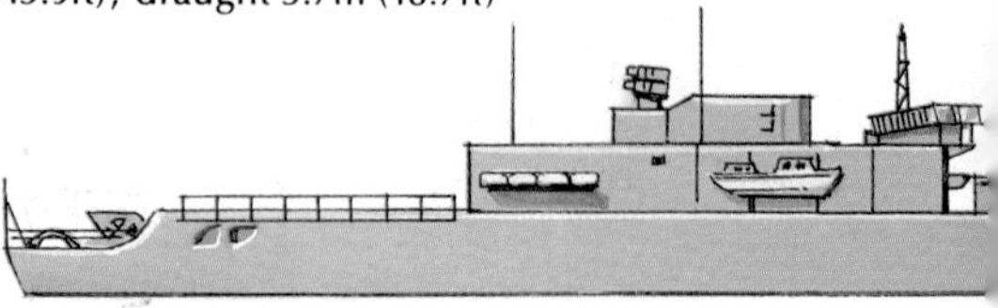

Gun Armament: One 100mm (3.9in) L/55 DP, and two 20mm (0.79in) AA
Missile Armament: Four container-launchers for eight MM38 (MM40 from D 642 onwards) Exocet surface-to-surface missiles, and one octuple launcher for 26 Crotale surface-to-air missiles
Anti-submarine Armament: Two tubes for 10 533mm (21in) L 5 torpedoes, and helicopter-launched Mk 46 A/S torpedoes
Aircraft: Two Westland Lynx Mk 2 helicopters
Propulsion: CODOG (COmbined Diesel Or Gas turbine) arrangement, with two Rolls-Royce Olympus TM3B gas turbines delivering 38,775kW (52,000bhp) to two shafts
Performance: Maximum speed 30kts on gas turbines or 21kts on diesels; range 17,600km (10,935mls) at

Complement: 15+90+111
Class
1. France (4+2+2)
GEORGES LEYGUES Commissioned December 1979
Remarks: Planned as a class of eight, the C70 ASW version will be divided into two four-ship sub-classes, the first four being C70/1 class ships as described above. The second quartet will be of the C70/2 type, with Flute ETBF towed-array sonar (in place of the DUBV 43 variable-depth sonar of the C70/1 type), an improved version of the Crotale Naval SAM system to provide more range and anti-missile capability, the 100mm (3.9in) Creusot-Loire Compact DP gun, and Vampir infra-red surveillance equipment.

CHARLES F. ADAMS CLASS
GUIDED-MISSILE DESTROYER

Country of Origin: USA
Displacement: 3,370 tons standard and 4,500 tons full load
Dimensions: Length 133.2m (437ft); beam 14.3m (47ft); draught 6.1m (20ft)

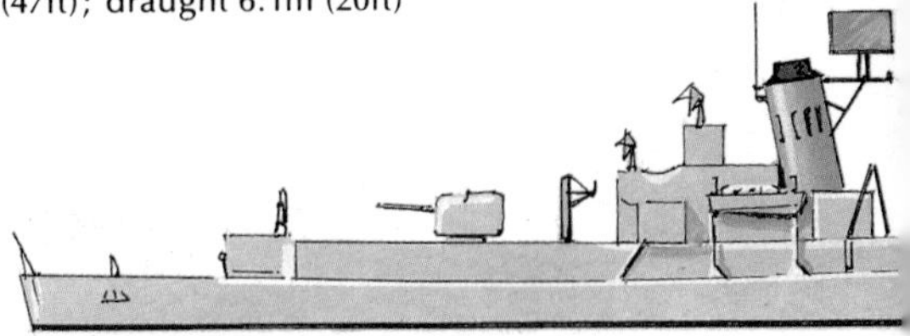

Missile Armament: Two quadruple launchers for eight RGM-84A Harpoon surface-to-surface missiles, and one Mk 11 twin launcher for 42 RIM-24 Tartar surface-to-air missiles
Anti-submarine Armament: One octuple launcher for RUR-5A ASROC missiles, and two triple Mk 32 tube mountings for 324mm (12.75in) Mk 46 A/S torpedoes
Propulsion: Four Babcock & Wilcox boilers supplying steam to two geared General Electric turbines delivering 52,200kW (70,000shp) to two shafts
Performance: Maximum speed 30kts; range 11,105km (6,900mls) at 14kts or 2,960km (1,840mls) at 30kts
Complement: 24+330
Class
1. USA (23)

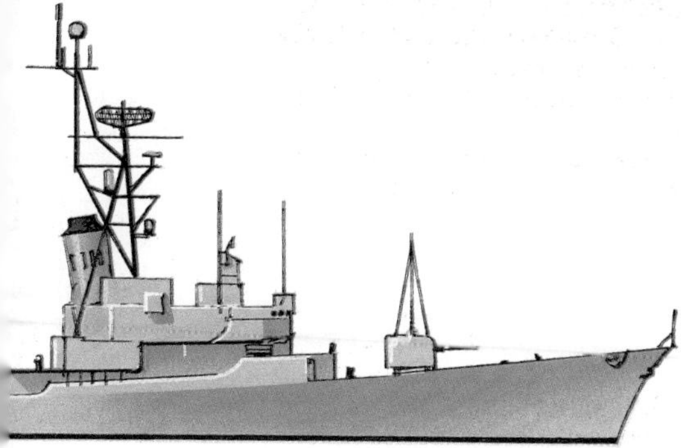

CHARLES F. ADAMS Commissioned September 1960
Remarks: Designed on the same basic hull as the Forrest Sherman class, and originally intended as continuations of the Hull class, the Charles F. Adams class was recast in the late 1950s as the US Navy's custom-built guided-missile destroyer class for the outer escort ring of fast carrier task groups. The first 13 ships were fitted with the Mk 11 twin launcher, later units having the Mk 13 single launcher. These are most capable ships, and are notable for their high levels of air-conditioned habitability. Though designed largely for area air-defence, the ships are also capable anti-submarine platforms.

COONTZ CLASS
GUIDED-MISSILE DESTROYER
Country of Origin: USA

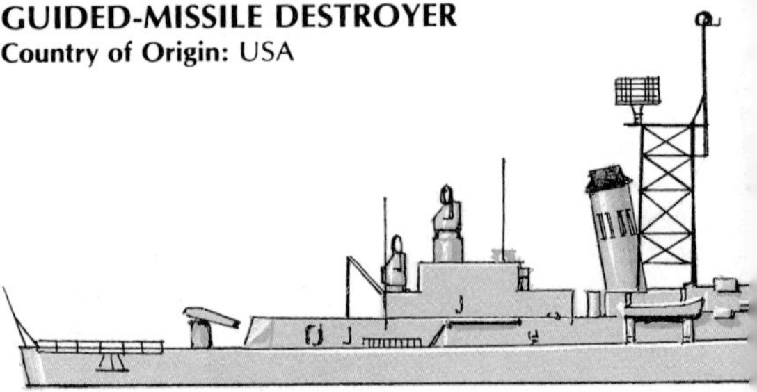

Displacement: 4,150-4,580 tons standard and 5,710-5,910 tons full load
Dimensions: Length 156.3m (512.5ft); beam 16.0m (52.5ft); draught 7.1m (23.4ft)
Gun Armament: One 127mm (5in) L/54 DP in a Mk 42 single mounting
Missile Armament: Two quadruple container-launchers for eight RGM-84A Harpoon surface-to-surface missiles, and one Mk 10 twin launcher for 40 RIM-66D Standard SM-2 surface-to-air missiles
Anti-submarine Armament: One octuple launcher for RUR-5A ASROC missiles, and two triple Mk 32 tube mountings for 324mm (12.75in) Mk 46 torpedoes
Aircraft: Provision for a helicopter on a platform aft
Propulsion: Four Babcock & Wilcox boilers supplying steam to two Allis-Chalmers geared turbines delivering 63,385kW (85,000shp) to two shafts

Performance: Maximum speed 33kts; range 9,255km (5,750mls) at 20kts
Complement: 21+356, plus provision for a flag staff of 7+12
Class
1. USA (10)
FARRAGUT Commissioned December 1960
Remarks: Though currently rated as guided-missile destroyers, and originally commissioned as guided-missile frigates, they have much of the capabilities of the US Navy's 'single-ended' guided-missile cruisers, with a primary missile armament of long-range Terrier surface-to-air missiles operated from a Mk 10 twin launcher in conjunction with two SPG-55B missile-control radars within the context of an operational scenario controlled through the ship's Naval Tactical Data System.

COUNTY CLASS
GUIDED-MISSILE DESTROYER
Country of Origin: UK

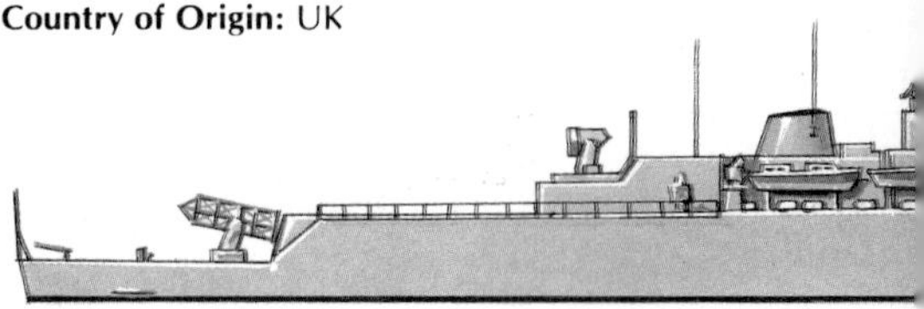

Displacement: 6,200 tons standard and 6,800 tons full load
Dimensions: Length 158.7m (520.5ft); beam 16.5m (54ft); draught 6.3m (20.5ft)
Gun Armament: Two 114mm (4.5in) L/45 DP in one Mk 6 twin mounting, and two 20mm (0.79in) Oerlikon AA in single mountings
Missile Armament: Four container-launchers for MM38 Exocet surface-to-surface missiles, one twin launcher for 36 Seaslug Mk 2 surface-to-air missiles, and two quadruple launchers for Sea Cat surface-to-air missiles
Anti-submarine Armament: Two triple 324mm (12.75in) STWS tube mountings for 12 Mk 46 A/S torpedoes
Aircraft: One Westland Wessex HAS Mk 3 helicopter
Propulsion: COSAG (COmbined Steam And Gas turbine) arrangement, with two Babcock & Wilcox boilers supplying steam to two sets of AEI geared turbines delivering 22,370kW (30,000shp) and four G6

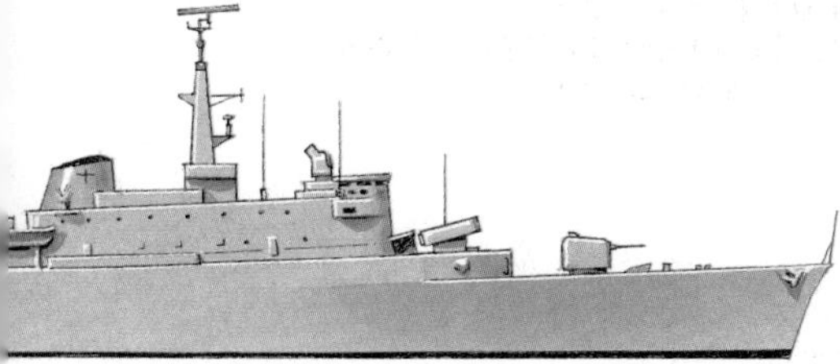

gas turbines delivering 22,370kW (30,000shp) to two shafts
Performance: Maximum speed 30kts; range 6,440km (4,000mls) at 28kts
Complement: 34+438
Class
1. Chile (1)
PRAT Commissioned March 1970
2. Pakistan (1)
BABUR Commissioned November 1963
3. UK (3)
ANTRIM Commissioned July 1970
Remarks: *Glamorgan* was hit by an Exocet missile on 12 June 1982 during the Falklands campaign the damage suffered being subsequently repaired. *Fife* was fitted with A/S torpedo tube and Lynx helicopter during lengthy refit, being recommissioned in March 1983. Apart from the two ships transferred to Chile and Pakistan, three other County class destroyers were paid off in the late 1970s.

D60 CLASS DESTROYER

Country of Origin: USA/Spain
Displacement: 2,425 tons standard and 3,480 tons full load

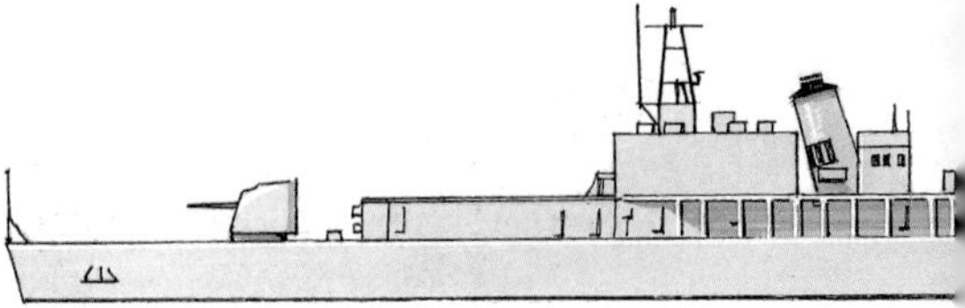

Dimensions: Length 119.0m (390.5ft); beam 12.4m (40.9ft); draught 5.8m (19ft)
Gun Armament: Four 127mm (5in) L/38 DP in two Mk 38 twin mountings
Anti-submarine Armament: One octuple launcher for RUR-5A ASROC missiles, and two triple Mk 32 tube mountings for 324mm (12.75in) Mk 44 torpedoes
Aircraft: One Hughes Model 500 helicopter
Electronics: One Lockheed SPS-40 air-search radar, one Raytheon/Sylvania SPS-10 surface-search radar, one navigation radar, one Western Electric Mk 25 or Western Electric Mk 28 radar-controlled gun fire-control system, one Mk 37 radar-controlled gun fire-control system, one Sangamo SQS-23 hull-mounted long-range active sonar used in conjunction with the Mk 114 A/S weapon-control system, and electronic support measures

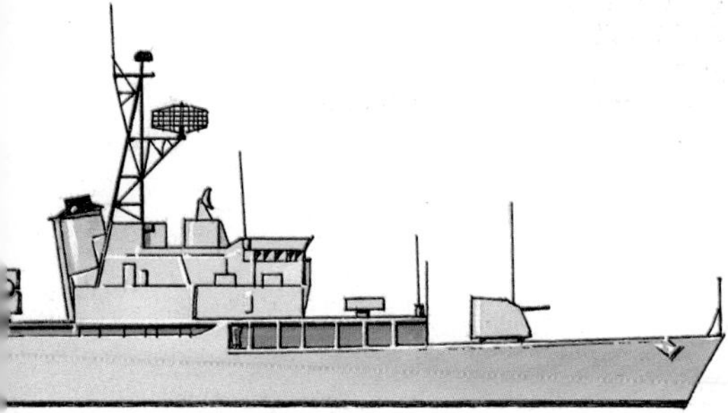

Propulsion: Four Babcock & Wilcox boilers supplying steam to two sets of General Electric or Westinghouse geared turbines delivering 44,740kW (60,000shp) to two shafts

Performance: Maximun speed 34kts; range 8,890km (5,525mls) at 15kts

Complement: 17+257

Class

1. Spain (5)

CHURRUCA Commissioned June 1945

Remarks: Operated under the Spanish designation 'D60', these Gearing (FRAM 1) class destroyers were bought from the USA in 1978 after being transferred in 1972 (first pair) and 1975 (last three). The *Blas de Lezo* (D65) differs from her sisters in having two forward gun mountings, no ASROC launcher, and torpedo tube mountings alongside the after funnel.

F67 CLASS
GUIDED-MISSILE DESTROYER
Country of Origin: France

Displacement: 4,580 tons standard and 5,745 tons full load
Dimensions: Length 1522.75m (501.1ft); beam 15.3m (50.2ft); draught 5.7m (18.7ft)
Gun Armament: Two 100mm (3.9in) L/55 DP in single mountings, and two 20mm (0.8in) AA in single mountings
Missile Armament: Six container-launchers for MM38 Exocet surface-to-surface missiles, and one octuple launcher for Crotale Naval surface-to-air missiles
Anti-submarine Armament: One launcher for 13 Malafon missiles, and two launchers for L 5 A/S torpedoes
Electronics: One Thompson-CSF DRBV 51 air and surface-search radar, one Thompson-CSF BRBV 26 air-search radar, one Thompson-CSF DRBC 32D fire-control radar used in conjunction with the Thompson-CSF Vega sensor and weapon-control sys-

tem, two Decca 1226 navigation radars, one CIT/ALCATEL DUBV 32 hull-mounted active search and attack sonar, one CIT/ALCATEL DUBV 43 variable-depth sonar, one SENIT 3 combat information sys-

tem, and two syllex chaff launchers (being replaced by CSEE Dagaie chaff launchers)

Propulsion: Four boilers supplying steam to two sets of Rateau double-reduction geared turbines delivering 40,565kW (54,400shp) to two shafts

Performance: Maximum speed 32kts; range 9,250km (5,750mls) at 18kts or 3,500km (2,175mls) at 30kts

Complement: 17+113+162

Class

1. France (3)

TOURVILLE Commissioned June 1974

Remarks: Potent anti-submarine destroyers with good anti-ship and anti-aircraft capabilities, the three F67 class destroyers serve with the French Atlantic Fleet, serving in turn as flagship of that force. The octuple Crotale Naval SAM launcher was fitted in place of the after 100mm (3.9in) gun during refits.

HAMBURG CLASS (TYPE 101A) GUIDED-MISSILE DESTROYER

Country of Origin: West Germany

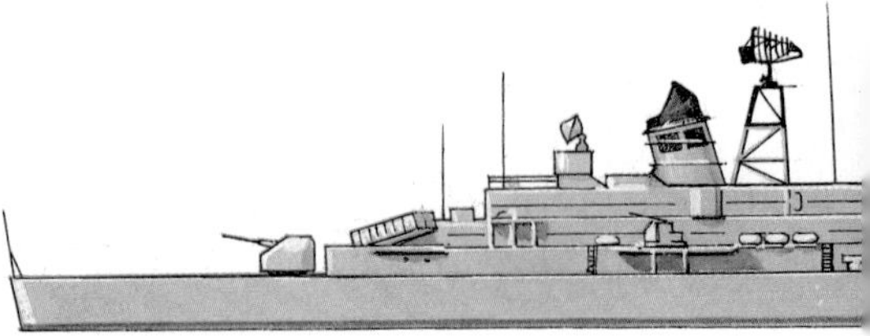

Displacement: 3,340 tons standard and 4,680 tons full load

Dimensions: Length 133.8m (439ft); beam 13.4m (44ft); draught 6.2m (20.3ft)

Gun Armament: Three 100m (3.9in) L/55 DP in single mountings, and eight 40mm (1.6in) Breda AA in four twin mountings

Missile Armament: Two twin container-launchers for MM38 Exocet surface-to-surface missiles

Anti-submarine Armament: Two four-barrel Bofors 375mm (14.75in) rocket-launchers, one quadruple 533mm (21 in) tube mounting for A/S torpedoes, and two depth-charge throwers

Propulsion: Four Wahodag boilers supplying steam to two sets of geared turbines delivering 50,710kW (68,000shp) to two shafts

Performance: Maximum speed 34kts; range 11,125km (6,915mls) at 13kts or 1,700km (1,055mls) at 34kts

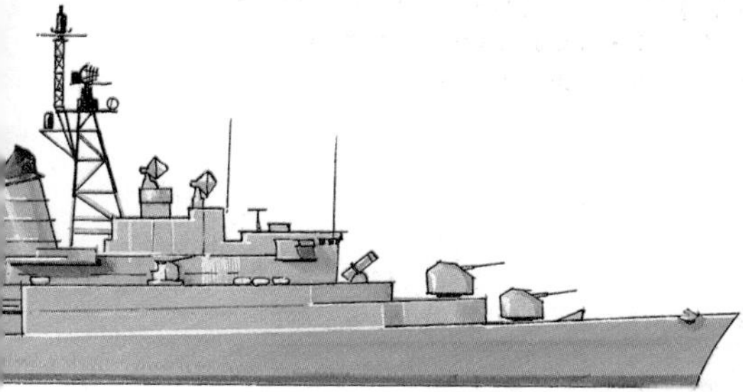

Complement: 19+249
Class
1. West Germany (4)
HAMBURG Commissioned March 1964
Remarks: Designed as general-purpose gun frigates optimized for anti-submarine warfare, the four units of the West German Hamburg class were modernized extensively during the 1970s: the 100mm (3.9in) gun in X position was replaced by two twin container-launchers for MM38 Exocet anti-ship missiles, the 40mm (1.6in) Bofors AA guns were replaced by four 40mm (1.6in) Breda twin AA mountings, a new Hollandse Signaalapparaten air-search radar was fitted, the five (three bow and two stern) fixed anti-ship torpedo tubes were removed, and two extra anti-submarine torpedo tubes were added. The West German navy hopes to keep these ships in service until the 1990s, which will entail further modification.

HARUNA CLASS
GUIDED-MISSILE DESTROYER

Country of Origin: Japan

Displacement: 4,700 tons standard and 6,300 tons full load

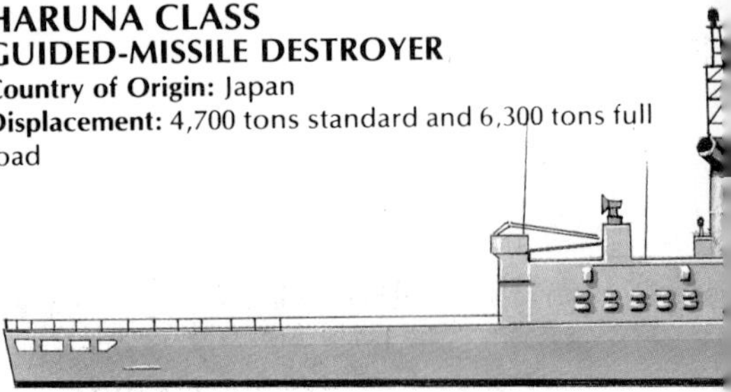

Dimensions: Length 153.0m (502ft); beam 17.5m (57.4ft); draught 5.1m (16.7ft)

Gun Armament: Two 127mm (5in) L/54 DP in two Mk 42 single mountings, and two 20mm (0.8in) Phalanx CIWS mountings

Missile Armament: Two quadruple container-launchers for eight RGM-84A Harpoon anti-ship missiles, and one octuple launcher for eight RIM-7 Sea Sparrow SAMS

Anti-submarine Armament: Two triple Mk 32 tube mountings for 324mm (12.75in) Mk 46 A/S torpedoes, one Mk 16 octuple launcher for 16 RUR-5A ASROC missiles, and helicopter-launched weapons

Aircraft: Three Sikorsky SH-3 Sea King helicopters in a hangar aft

Electronics: One Hughes SPS-52 radar, one OPS-17 surface-search radar, two Type 72 gun fire-control radars, one Hollandse Signaalapparaten WM-25 mis-

sile-control radar, TACAN, one ESM suite, one OQS-3 hull sonar, and one SQS-3S(J) variable-depth sonar
Propulsion: Boilers supplying steam to two sets of geared turbines delivering 52,200kW (70,000shp) to two shafts
Performance: Maximum speed 32kts
Complement: 364
Class
1. Japan (2)
HARUNA Commissioned February 1973
HIEI Commissioned November 1974
Remarks: Key features of the *Haruna* are the large helicopter platform and associated hangar (large enough to handle three Mitsubishi-Sikorsky SH-3 anti-submarine helicopters), and the octuple ASROC launcher just forward of the bridge structure, together with two 127mm (5in) DP gun mountings.

HULL CLASS DESTROYER

Country of Origin: USA
Displacement: 2,800 to 3,000 tons standard and 3,960 to 4,200 tons full load

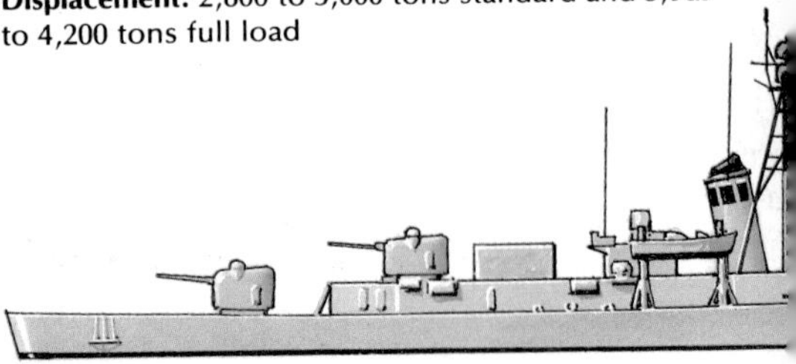

Dimensions: Length 127.4m (418ft); beam 13.7m (45ft); draught 7.0m (23ft)
Gun Armament: Three 127mm (5in) L/54 DP in three Mk 24 single mountings
Anti-submarine Armament: Two triple Mk 32 tube mountings for 324mm (12.75in) Mk 46 A/S torpedoes
Propulsion: Four Babcock & Wilcox boilers supplying steam to two sets of General Electric geared turbines delivering 52,200kW (70,000shp) to two shafts
Performance: Maximum speed 33kts; range 8,370km (5,200mls) at 20kts
Complement: 17+275, or (A/S modified vessels) 17+287
Class
1. USA (5)
HULL Commissioned July 1958

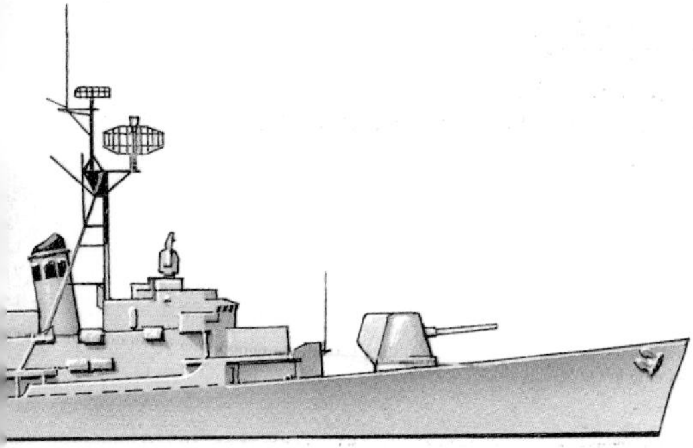

Remarks: The Hull class is almost identical with the Forrest Sherman class, but was designated separately because of its modified bow. Together with the Forrest Sherman class, these ships were the first US warships to have more firepower aft than forward. The *Hull* was used as trials ship for the 203mm (8in) Major Caliber Light Weight Gun, mounted in place of the forward 127mm (5in) mounting. The trials confirmed the overall value of the MCLWG, especially for enhanced shore-bombardment capability, but the original 127mm (5in) mounting was replaced in February 1979 with the cancellation of the MCLWG programme for financial reasons. With the exception of the *Edson*, which is used for engine-room training with the Naval Reserve Force, all destroyers of the Hull class have been placed in reserve.

IMPAVIDO CLASS
GUIDED-MISSILE DESTROYER

Country of Origin: Italy
Displacement: 3,200 tons standard and 3,990 tons full load

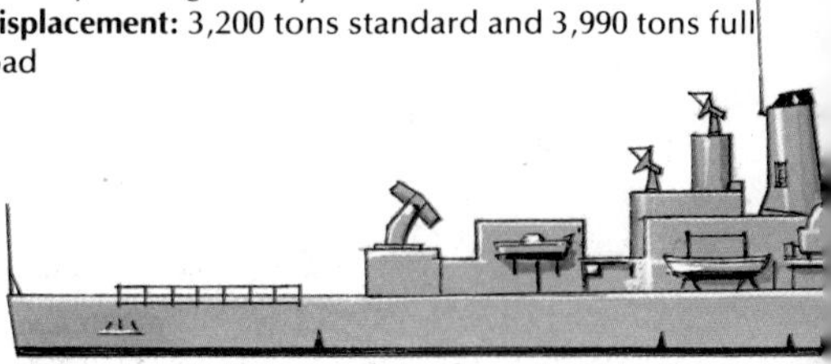

Dimensions: Length 131.3m (429.5ft); beam 13.6m (44.7ft); draught 4.5m (14.8ft)
Gun Armament: Two 127mm (5in) L/38 DP in one Mk 38 twin mounting, and four 76mm (3in) L/62 DP in single mountings
Missile Armament: One Mk 13 single launcher for 40 RIM-66 Standard surface-to-air missiles
Anti-submarine Armament: Two triple Mk 32 tube mountings for 324mm (12.75in) Mk 46 A/S torpedoes
Aircraft: Provision for a light helicopter on the platform aft
Electronics: One RCA SPS-12 air-search radar, one Hughes SPS-52B 3D air-search radar, one SMA SPQ 2 surface-search radar, two SPG-51B missile-control radars used in conjunction withtwo Raytheon Mk 73 missile fire-control systems. three Selenia Orion RTN 10X gun-control radars used in conjunction with the

three ELSAG Argo NA10 gun fire-control systems, one SQS-23 hull-mounted sonar, one ESM suite and two Breda SCLAR 20-barrel chaff launcher systems

Propulsion: Four Foster-Wheeler boilers supplying steam to two sets of Tosi geared turbines delivering 52,200kW (70,000shp) to two shafts

Performance: Maximum speed 33kts; range 6,100km (3,790mls) at 20kts of 2,775km (1,725mls) at 30kts

Complement: 23+317

Class

1. Italy (2)

IMPAVIDO Commissioned November 1963

Remarks: Precursors to the Audace class, the two Impavido class guided missile destroyers are capable though limited anti-aircraft ships, but lack adequate anti-submarine capability as they have no embarked helicopter, or even provision for such a helicopter.

IROQUOIS CLASS
DESTROYER

Country of Origin: Canada
Displacement: 3,550 tons standard and 4,700 tons full load

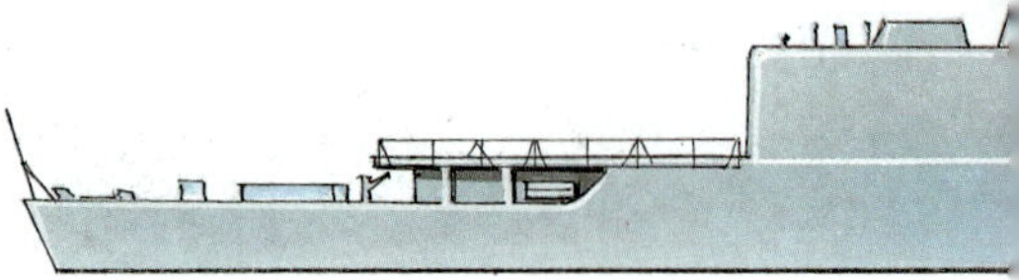

Dimensions: Length 129.8m (426ft); beam 15.2m (50ft); draught 4.7m (15.5ft)
Gun Armament: One 127mm (5in) L/54 DP OTO-Melara Compact
Missile Armament: Two Raytheon quadruple launchers for 32 RIM-7 Sea Sparrow surface-to-air missiles
Anti-submarine Armament: Two triple Mk 32 tube mountings for 324mm (12.75in) Mk 46 torpedoes, one Mk NG 10 Limbo mortar, and helicopter-launched weapons
Aircraft: Two Sikorsky CH-124 Sea King helicopters in a hangar amidships
Propulsion: COGOG (Combined Gas turbine Or Gas turbine) arrangement, with two Pratt & Whitney FT4A2 gas turbines delivering 37,285kW (50,000shp) or two Pratt & Whitney FT12AH3 gas turbines delivering 5,520kW (7,400shp) to two shafts

Performance: Maximum speed 29+kts on main engines or 18kts on cruising engines, range 8,370km (5,200mls) at 20kts
Complement: 20+225, plus an air unit of 7+33
Class
1. Canada (4)
IROQUOIS Commissioned July 1972
Remarks: Until the arrival of a new destroyer class (to permit retirement of the St Laurent class frigates), the four Iroquois class ships are the only destroyers in Canadian service, and are designed primarily for an anti-submarine role, with helicopters as the basic weapon/sensor system in concert with limited shipborne weapons and good shipborne sensors. The helicopters are flown in conjunction with haul-down and deck-control systems to maximize the possibility of continued operations even under adverse weather conditions.

KANIN CLASS
GUIDED-MISSILE DESTROYER

Country of Origin: USSR
Displacement: 3,700 tons standard and 4,700 tons full load

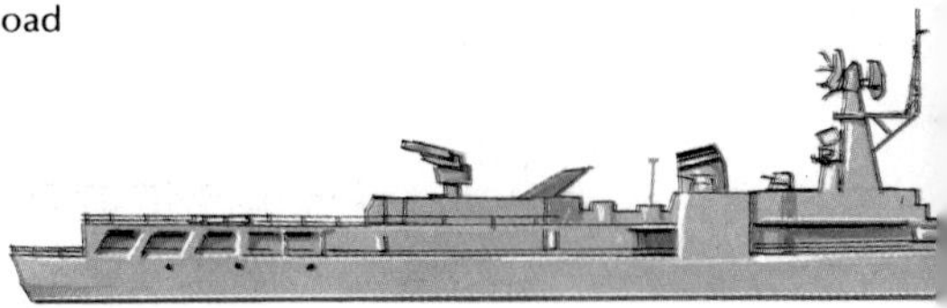

Dimensions: Length 139.0m (455.9ft); beam 14.7m (48.2ft); draught 5.0m (16.4ft)
Gun Armament: Eight 57mm (2.25in) L/70 AA in two quadruple mountings, and eight 30mm (1.2in) AA in four twin mountings
Missile Armament: One twin launcher for 22 SA-N-1 'Goa' surface-to-air missiles
Torpedo Armament: Two quintuple 533mm (21in) AS/ASW tube mountings.
Anti-submarine Armament: Three RBU 6000 12-barrel rocket-launchers
Aircraft: Provision for one helicopter on a platform aft
Propulsion: Four boilers supplying steam to two sets of geared turbines delivering 59,655kW (80,000shp) to two shafts
Performance: Maximum speed 34kts; range 8,350km (5,190mls) at 16kts or 2,050km (1,275mls) at 33kts
Complement: 350

Class
1. USSR (8)
BOYKY; DERZKY; GNEVNY; GORDY; GREMYASHCHY; UPORNY; ZHGUCHY; ZORKY

Remarks: The Kanin class anti-aircraft destroyers were produced by a programme of modification from eight Krupny class anti-ship destroyers, the alterations including the deletion of the two SS-N-1 anti-ship missile launchers and their replacement by a single twin launcher (aft) and an additional RBU 6000 anti-submarine rocket-launcher (foredeck), the enlargement of the helicopter platform, a different and longer bow (probably for a new bow-mounted sonar), the halving of the 57mm (2.25in) armament, the enlarging of the new bridge structure, the addition of new radars and, later, the addition of four 30mm (1.18in) twin AA mountings.

KASHIN CLASS
GUIDED-MISSILE DESTROYER
Country of Origin: USSR

Displacement: 3,750 tons standard and 4,500 tons full load
Dimensions: Length 143.3m (470.7ft); beam 15.8m (51.8ft); draught 4.7m (15.4ft)
Gun Armament: Four 76mm (3in) L/60 DP in two twin mountings
Missile Armament: Two twin launchers for 32 SA-N-1 'Goa' surface-to-air missiles
Torpedo Armament: One quintuple 533mm (21in) AS/ASW tube mounting
Anti-submarine Armament: Two RBU 6000 12-barrel rocket-launchers, and two RBU 1000 six-barrel rocket-launchers
Propulsion: Four gas turbines delivering 70,095kW (94,000hp) to two shafts
Performance: Maximum speed 35kts; range 8,350km (5,190mls) at 18kts of 2,600km (1,615mls) at 34kts
Complement: 280

Class
1. USSR (13)
KOMSOMOLETS UKRAINY; KRASNY-KAVKAZ; KRASNY-KRIM; OBRAZTOVY; ODARENNY; PROVORNY; RESHITELNY; SKORY; SMETLIVY; SOOBRAZITELNY; SPOSOBNY; STEREGUSHCHY; STROGY
Remarks: The Kashin class anti-aircraft guided-missile destroyer was the world's first major warship designed with all-gas turbine propulsion, and was possibly intended for AA escort of the contemporary Kinda class anti-ship cruisers. The Kashin (Modified) class programme was started in 1972 to provide some of the ships with improved anti-ship and anti-submarine capabilities by the lengthening of the hull to make possible the accommodation of SS-N-2C anti-ship missiles, a helicopter platform and variable depth sonar. The Indian Kashin II class is a further development of the Kashin (Modified) class.

KIDD CLASS
GUIDED-MISSILE DESTROYER

Country of Origin: USA

Displacement: 6,210 tons standard and 9,200 tons full load
Dimensions: Length 171.6m (563ft); beam 16.8m (55ft); draught 9.1m (30ft) to sonar dome
Gun Armament: Two 127mm (5in) L/54 DP in two Mk 45 single mountings, and two 20mm (0.79in) Phalnax Mk 15 close-in weapon system mountings
Missile Armament: Two quadruple container-launchers for eight RGM-84A Harpoon surface-to-surface missiles, and two Mk 26 twin launchers for 52 RIM-67A Standard-ER surface-to-air missiles
Anti-submarine Armament: Two triple Mk 32 tube mountings for 324mm (12.75in) Mk 46 A/S torpedoes, 16 RUR-5A ASROC missiles (fired from the Mk 26 launchers), and helicopter-launched weapons
Aircraft: Two Kaman SH-2F Seasprite helicopters
Propulsion: Four General Electric LM 2500 gas turbines delivering 59,655kW (80,000shp) to two shafts

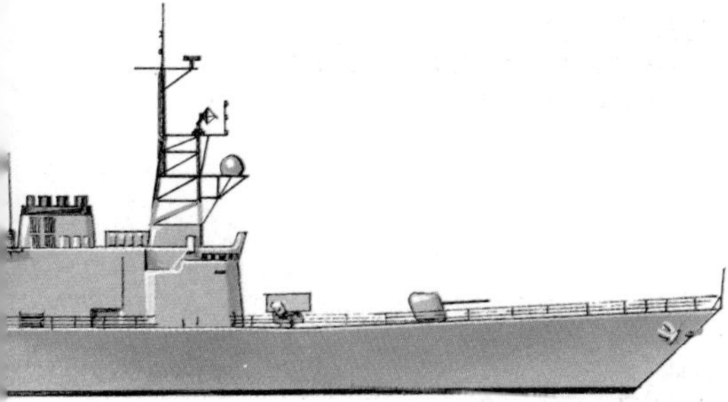

Performance: Maximum speed 33kts; range 14,805km (9,200mls) at 17kts or 6,115km (3,800mls) at 30kts
Complement: 20+318
Class
1. USA (4)
KIDD Commissioned May 1981
CALLAGHAN Commissioned August 1981
SCOTT Commissioned October 1981
CHANDLER Commissioned March 1982
Remarks: Though rated as guided-missile destroyers, the four ships of the Kidd class are more akin to 'double-ended' guided-missile cruisers in terms of performance, capability and size. There is little doubt that they are the most powerful ships of their type in the world, combining excellent anti-ship, anti-submarine and anti-aircraft capabilities (both sensors and weapons) in a design of first-class habitability.

MEKO 360 CLASS GUIDED-MISSILE DESTROYER

Country of Origin: West Germany/Nigeria
Displacement: 3,630 tons full load

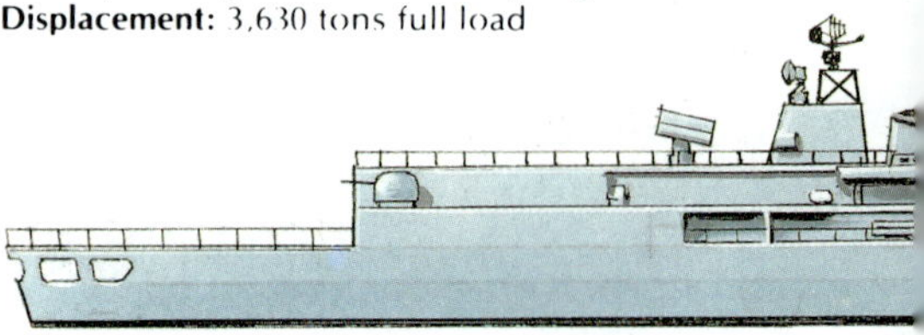

Dimensions: Length 125.6m (412ft); beam 15.0m (49.2ft); draught 4.3m (14.1ft)
Gun Armament: One 127mm (5in) L/54 DP OTO-Melara Compact, and eight 40mm (1.6in) Breda L/70 AA in four twin mountings
Missile Armament: Two quadruple container-launchers for eight Otomat surface-to-surface missiles, and one octuple launcher for 24 Aspide surface-to-air missiles
Anti-submarine Armament: Two triple Plessey tube mountings for 324mm (12.75in) A 244S A/S torpedoes, and helicopter-launched weapons
Aircraft: One helicopter in a hanger aft
Propulsion: CODOG (COmbined Diesel Or Gas turbine) arrangement, with two MTU 20V956 TB92 diesels delivering 7,455kW (10,000hp) and two Rolls-Royce Olympus TM3B gas turbines delivering 41,760kW (56,000hp) to two shafts

Fuel: Diesel oil and kerosene
Performance: Maximum speed 30.5kts on gas turbines; range 12,000km (7,455mls) at cruising speed
Complement: 200
Class
1. Argentina (4)
LA ARGENTINA Commissioned 1983
2. Nigeria (1)
ARADU Commissioned September 1981
Remarks: The Meko 360 class is the first fully modular design for a major warship in the world, the modular design of the armament, sensors and other key systems making replacement and updating a relatively simple matter of component replacement. The four Argentinian ships differ somewhat from the earlier Nigerian ship, having a gas-turbine propulsion system (two Rolls-Royce Olympus TM1Bs and two Rolls-Royce Tyne RM1Cs).

MINEGUMO CLASS
DESTROYER

Country of Origin: Japan

Displacement: 2,050 tons standard and 2,150 tons full load

Dimensions: Length 114.9m (377ft); beam 11.8m (38.7ft; draught 4.0m (13.1ft)

Gun Armament: Four 76mm (3in) L/50 DP in two Mk 33 twin mountings

Anti-submarine Armament: Two triple Type 68 tube mountings for 324mm (12.75in) Mk 44/46 A/S torpedoes, one Mk 16 octuple launcher for RUR-5A ASROC missiles, and one Bofors Type 71 four-barrel rocket-launcher

Electronics: One OPS-11 air-search radar, one OPS-17 surface-search radar, two Mk 35 gun-control radars used in conjunction with Mk 56 and Mk 63 gun fire-control systems, one OQS-3 hull-mounted sonar

Propulsion: Six Mitsubishi diesels delivering 19,760kW (26,500bhp) to two shafts
Performance: Maximum speed 27kts; range 12,975km (8,065 mls) at 20kts
Complement: 210
Class
1. Japan (3)
Name:
MINEGUMO Commissioned August 1968
NATSUGUMO Commissioned April 1969
MURAKUMO Commissioned August 1970
Remarks: The Minegumo class was designed to the same basic pattern as the Yamagumo class, with provision for DASH drone helicopters. This anti-submarine weapon system proved to be a failure and in 1978 the *Murakumo* was rearmed.

SHEFFIELD CLASS (TYPE 42)
GUIDED-MISSILE DESTROYER

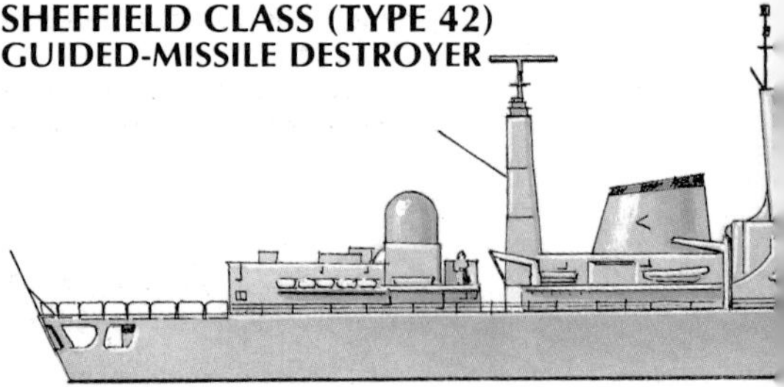

Country of Origin: UK
Displacement: 3,850 tons standard and 4,350 tons full load
Dimensions: Length 125.6m (412ft); beam 14.3m (47ft); draught 5.8m (19ft) to screws
Gun Armament: One 114mm (4.5in) L/55 DP in one Mk 8 single mounting, four 20mm (0.79in) Oerlikon AA, and two twin 30mm (1.18in) AA
Missile Armament: One twin launcher for 24 Sea Dart surface-to-air and surface-to-surface missiles
Anti-submarine Armament: Two triple tube mountings for 324mm (12.75in) Mk 46 A/S torpedoes, and helicopter-launched weapons
Aircraft: One Westland Lynx HAS.Mk 2 helicopter in a hangar aft
Propulsion: COGOG (COmbined Gas turbine Or Gas Turbine) arrangement, with two Rolls-Royce Tyne RM1A gas turbines delivering 6,340kW (8,500shp) and

two Rolls-Royce Olympus TM3B gas turbines delivering 41,760kW (56,000shp) to two
Performance: Maximum speed 29kts on Olympus turbines or 18kts on Tyne turbines; range 7,400km

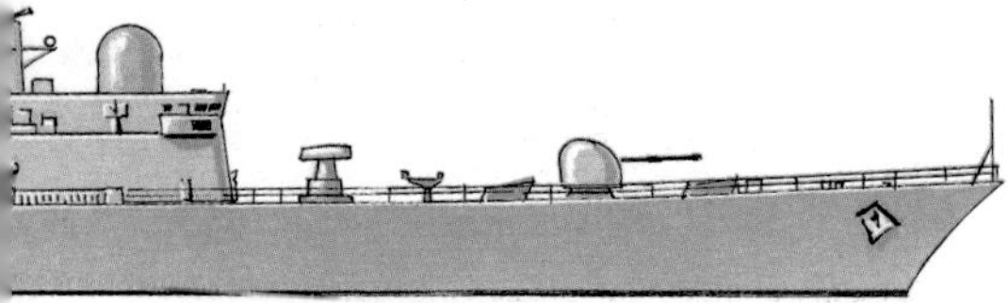

(4,600mls) at 18kts or 1,205km (750mls) at 29kts
Complement: 24+229, with a maximum accommodation of 312 possible
Class
1. **UK** (12)
Name:
BIRMINGHAM Commissioned December 1976
Remarks: Designed for the area defence of a Royal Navy task group, the Type 42 or Sheffield class guided-missile destroyer was extensively blooded in the Falklands campaign of 1982, and found wanting in several important facets, notably damage control and point defence against sea-skimmer missiles and low-level aircraft. The *Sheffield* and *Coventry* of this class were lost in the Falklands war, and urgent steps have been taken to improve the most serious defects, notably the addition of two twin 30mm (1.18in) AA and two single 20mm (0.79in) AA mountings at the cost of a reduction in boat strength.

SHIRANE CLASS DESTROYER

Country of Origin: Japan
Displacement: 5,200 tons standard and 6,800 tons full load

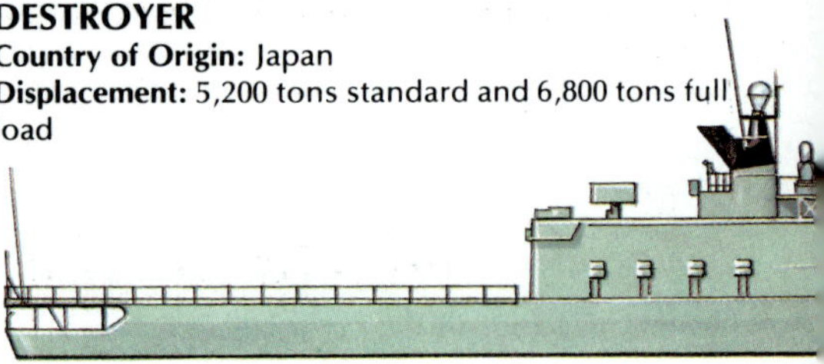

Dimensions: Length 159.0m (521.5ft); beam 17.5m (57.5ft); draught 5.3m (17.5ft)
Gun Armament: Two 127mm (5in) L/54 DP in two Mk 42 single mountings, and two 20mm (0.79in) Phalanx close-in weapon system mountings
Missile Armament: Two quadruple container-launches for eight RIM-7 Sea Sparrow surface-to-air missiles
Anti-submarine Armament: Two triple Type 68 tube mountings for 324mm (12.75in) Mk 44/46 torpedoes, one Mk 16 octuple launcher for 16 RUR-5A ASROC missiles, and helicopter-launched weapons
Aircraft: Three Sikorsky SH-3 Sea King helicopters in a hangar amidships
Electronics: One Hughes SPS-52B 3D search radar, one OPS-28 surface-search radar, two Type 72 gun-control radars, one Hollandse Signaalapparaten WM-25 missile-control radar, TACAN, one ESM suite,

one OQS-101 hull-mounted sonar, one EDO SQS-35(J) variable-depth sonar, and one SQR-18 passive towed sonar

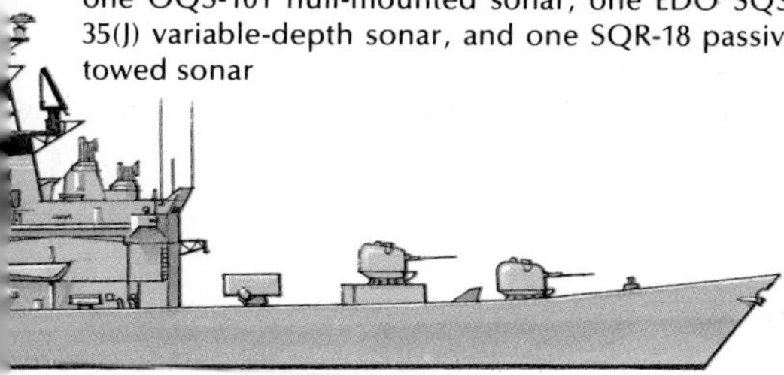

Propulsion: boilers supplying steam to two sets of geared turbines delivering 52,200kW (70,000shp) to two shafts
Performance: Maximum speed 32kts
Complement: 350
Class
1. Japan (2)
SHIRANE Commissioned March 1980
KURAMA Commissioned March 1981
Remarks: The two 'Shirane' class ships are the most modern destroyers in service with the Japanese Maritime Self-Defence Force, and are optimized for anti-submarine warfare, with ASROC and three helicopters for offensive operations, and two triple Type 68 tube mountings for self-defence. The design is a modification of that of the Haruna class. Of the two funnels, the forward unit is set slightly to port and the after unit slightly to starboard of the centreline.

SPRUANCE CLASS GUIDED-MISSILE DESTROYER

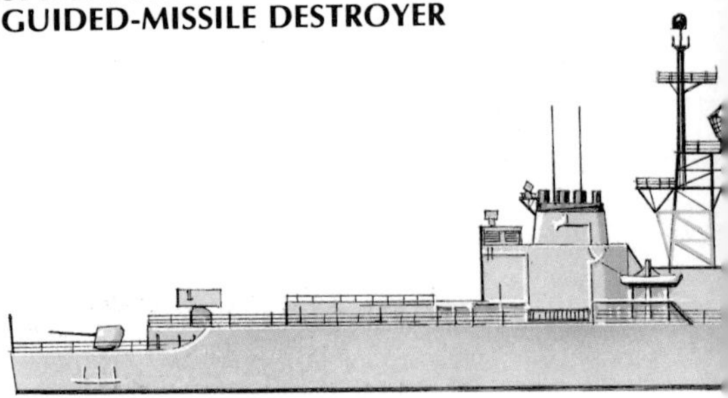

Country of Origin: USA
Displacement: 5,830 tons standard and 7,810 tons full load
Dimensions: Length 171.7m (563.2ft); beam 16.8m (55.1ft); draught 8.8m (29ft) to sonar dome
Gun Armament: Two 127mm (5in) L/54 DP in two Mk 45 single mountings, and two 20mm (0.79in) Phalanx Mk 15 close-in weapon system mountings
Missile Armament: Two quadruple container-launchers for eight RGM-84A Harpoon surface-to-surface missiles, two quadruple container-launchers for eight BGM-109 Tomahawk surface-to-surface missiles and one Mk 29 octuple launcher for 24 RIM-7 Sea Sparrow surface-to-air missiles
Anti-submarine Armament: Two triple Mk 32 tube mountings for 14 324mm (12.75in) Mk 46 A/S tor-

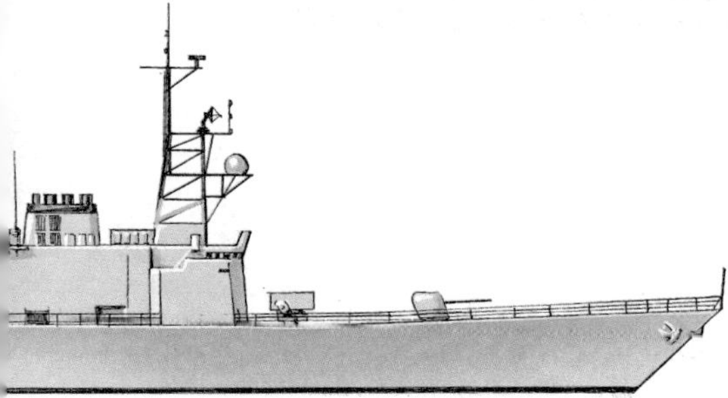

pedoes, and one octuple launcher for 24 RUR-5A ASROC missiles

Aircraft: One Sikorsky SH-3 Sea King or two Kaman SH-2 Seasprite helicopters in a hangar aft

Propulsion: Four General Electric LM 2500 gas turbines delivering 59,655kW (80,000 shp) to two shafts

Performance: Maximum speed 33kts; range 11,105km (6,900 mls), at 20kts

Complement: 24+272

Class

1. USA (31+1)

SPRUANCE Commissioned September 1975

Remarks: The US Navy's most important anti-submarine destroyer force, the Spruance class destroyers are designed for the escort of carrier task forces in conjunction with anti-aircraft cruisers.

SUFFREN CLASS
GUIDED-MISSILE DESTROYER

Country of Origin: France
Displacement: 5,090 tons standard and 6,090 tons full load

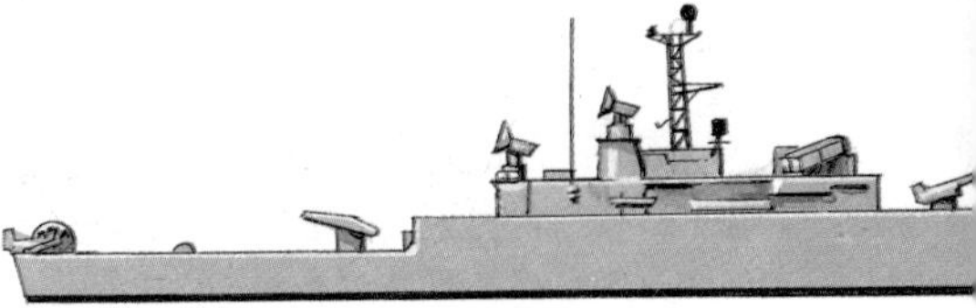

Dimensions: Length 157.6m (517.1ft); beam 15.54m (51ft); draught 6.1m (20ft)
Gun Armament: Two 100mm (3.9in) L/55 DP in two single mountings, and four 20mm AA in single mountings
Missile Armament: Four container-launchers for four MM.38 Exocet surface-to-surface missiles, and one twin launcher for 48 Masurca surface-to-air missiles
Anti-submarine Armament: One single launcher for 13 Malafon missiles, and four launchers for L 5 A/S torpedoes
Electronics: One Thomson-CSF DRBN 32 navigation radar, two Thomson-CSF DRBR 52 SAM-control radars, one Thomson-CSF DRBC 32A gun-control radar, one CIT/Alcatel DUBV 23 hull-mounted active search and attack sonar, one CIT/Alcatel DUBV 43 variable-depth sonar, one SENIT 1 combat infor-

mation system, TACAN, one ESM system, and two Dagaie chaff launchers
Propulsion: Four boilers supplying steam to two sets

of Rateau double-reduction geared turbines delivering 54,065kW (72,500shp) to two shafts
Performance: Maximum speed 34kts; range 9,500km (5,905 mls) at 18kts or 4,450km (2,765 mls) at 29kts
Complement: 23+332
Class
1. France (2)
SUFFREN Commissioned July 1967
DUQUESNE Commissioned April 1970
Remarks: This class is designed for the area protection of France's two aircraft carriers against air and submarine threats. Forward are two 100mm (3.9in) DP guns with their associated DRBC 32A fire-control radar and domed DRBI 23 surveillance radar; amidships is the Malafon anti-submarine launcher with aircraft-type missile; and aft is the Masurca twin launcher with DRBR 51 radars.

T56 CLASS DESTROYER

Country of Origin: France

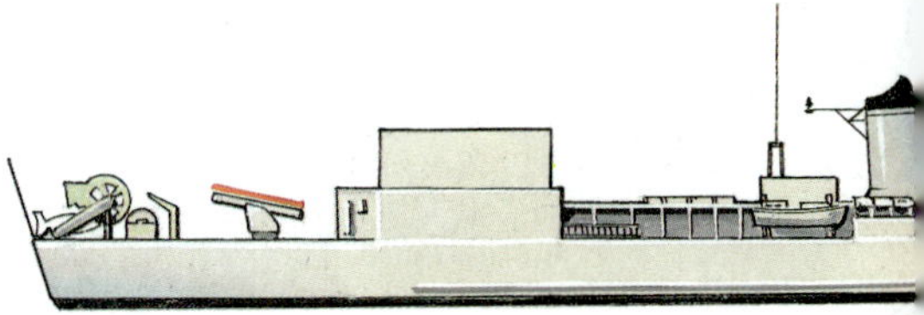

Displacement: 2,750 tons standard and 3,910 tons full load

Dimensions: Length 132.8m (435.7ft); beam 12.7m (41.7ft); draught 5.4m (17.7ft)

Gun Armament: Two 100mm (3.9in) L/55 DP in two single mountings

Anti-submarine Armament: One launcher for Malafon missiles, two triple 550mm (21.7in) tube mountings for K 2 and L3 A/S torpedoes, and helicopter-launched weapons

Aircraft: One Aerospatiale Alouette III helicopter in a hangar aft

Electronics: One Thomson-CSF DRBV 50 combined air-and surface-search radar, one Thomson-CSR DRBV 22 air-search radar, one Thomson-CSF DRBC 32A gun-control radar, one Thomson-CSR BRBN 32 navigation radar, one CIT/ALCATEL DUBV 23 hull-mounted active search and attack sonar, one CIT/

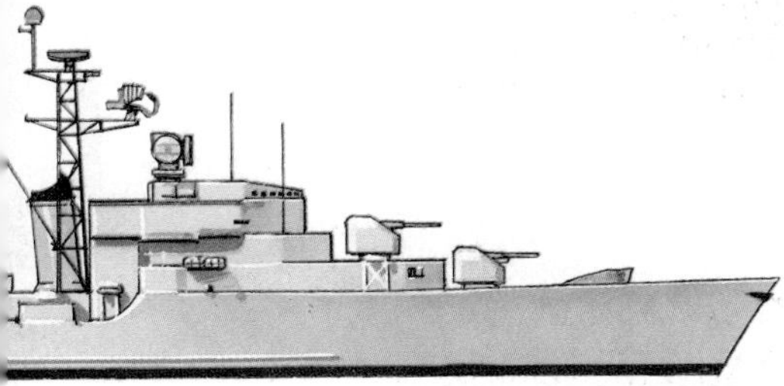

ALCATEL DUBV 43 variable-depth sonar, URN-22 TACAN, and electronic warfare systems
Propulsion: Four Indret boilers supplying steam to two sets of Rateau geared turbines delivering 46,980kW (63,000shp) to two shafts
Performance: Maximum speed 34kts; range 9,250km (4,750 mls) at 18kts
Complement: 15+255
Class
1. France (1)
Name:
LA GALISSONNIÈRE: Commissioned July 1962
Remarks: With the same hull and machinery as the T 47 and T53 classes, the sole T56 class ship, *La Galissonnière* was built as an experimental anti-submarine ship, and was the first destroyer armed with the Malafon missile system. The sides of the hangar open outwards to form the helicopter platform.

TROMP CLASS
GUIDED-MISSILE DESTROYER

Country of Origin: Netherlands

Displacement: 3,900 tons standard and 4,580 tons full load
Dimensions: Length 138.2m (453.3ft); beam 14.8m (48.6ft); draught 4.6m (15.1ft)
Gun Armament: Two 120mm (4.7in) L/50 DP in one twin mounting
Missile Armament: Eight container-launchers for 16 RGM-84A Harpoon surface-to-surface missiles, one Mk 13 single launcher for 40 RIM-24B Tartar surface-to-air missiles, and one octuple launcher for 16 RIM-7 Sea Sparrow surface-to-air missiles
Anti-submarine Armament: Two triple Mk 32 tube mountings for 324mm (12.75in) Mk 46 A/S torpedoes
Aircraft: One Westland Lynx helicopter in a hangar aft
Electronics: One Hollandse Signaalapparaten 3D-MITR 3D radar, one Hollandse Signaalapparaten WM-25 surface-search, gun-control and Sea Sparrow-control radar, two Raytheon SPG-51 Tartar-control radars, two Decca navigation radars, one CWE

610 hull-mounted sonar, one SEWACO I action-data system, one Daisy data-handling and underwater weapon fire-control system, and two Knebworth Corvus chaff launchers

Propulsion: COGOG (COmbined Gas turbine Or Gas turbine) arrangement, with two Rolls-Royce Tyne TM1 gas turbines delivering 5,965kW (8,000hp) and two Rolls-Royce Olympus TM3B gas turbines delivering 37,285kW (50,000hp) to two shafts

Performance: Maximum speed 30kts; range 9,250km (5,750 mls) at 18kts

Complement: 34+267

Class

1. Netherlands (2)

TROMP Commissioned October 1975

Remarks: The Dutch guided-missile destroyer *De Ruyter* provides anti-aircraft defence for the Royal Dutch Navy's ASW Group II operating in the eastern Atlantic under NATO's auspices.

VOSPER THORNYCROFT MK 10 CLASS
GUIDED-MISSILE DESTROYER

Country of Origin: UK/Brazil

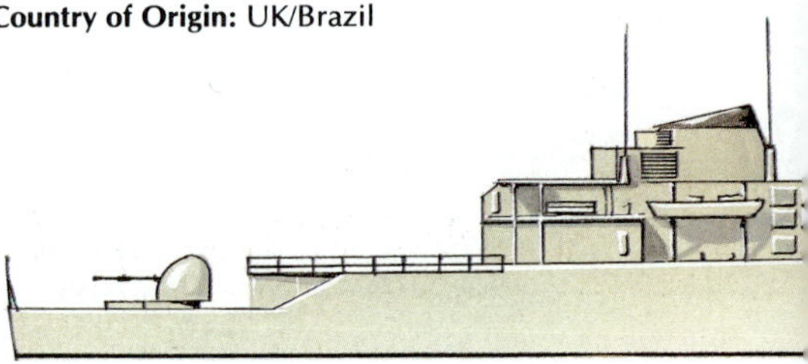

Displacement: 3,200 tons standard and 3,800 tons full load

Dimensions: Length 129.2m (424ft); beam 13.5m (44.2ft); draught 5.5m (18.2ft)

Gun Armament: One 114mm (4.5in) L/55 DP in Vickers Mk 8 single mountings, and two 40mm (1.58in) Bofors L/70 AA

Missile Armament: Two twin container-launchers for four MM38 Exocet surface-to-surface missiles (Fs 42 and 43), and two triple launchers for Sea Cat surface-to-air missiles

Anti-submarine Armament: One 375mm (14.76in) Bofors two-barrel rocket-launcher, two triple Plessey STWS 1 tube mountings for 324mm (12.75in) Mk 44/46 A/S torpedoes, one depth-charge rail, and one launcher for 10 Ikara missiles, and helicopter-launched weapons

Aircraft: One Westland Lynx helicopter
Propulsion: CODOG (COmbined Diesel Or Gas turbine) arrangement, with four MTU diesels delivering 11,750kW (15,760shp) or two Rolls-Royce Olympus TM3B gas turbines delivering 41,760kW (56,000shp) to two shafts
Performance: Maximum speed 30kts on gas turbines or 22kts on diesels; range 9,825km (6,105 mls) at 17kts on two diesels, or 7,775km (4,830mls) at 19kts on four diesels, or 2,400km (1,490mls) at 28kts on gas turbines; endurance d45 days
Complement: 21+179
Class
1. Brazil (6 'Niteroi' class)
NITEROI Commissioned November 1976
Remarks: The first four craft in the class were built by Vosper Thornycroft Ltd.

A69 CLASS
GUIDED-MISSILE FRIGATE

Country of Origin: France

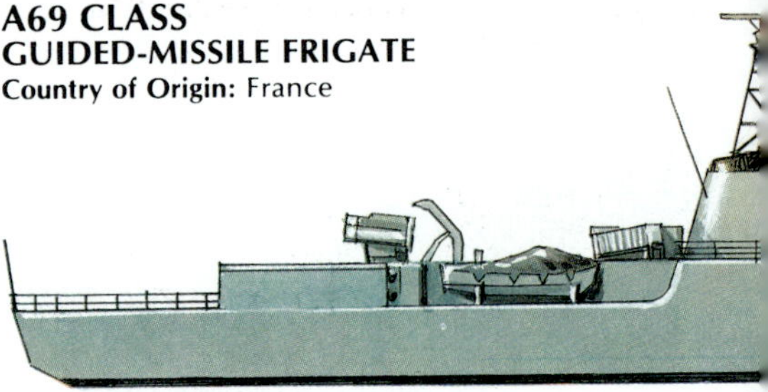

Displacement: 950 tons standard and 1,170 tons (or 1,250 tons in later ships) full load
Dimensions: Length 80.0m (262.5ft); beam 10.3m (33.8ft); draught 5.3m (17.4ft) to sonar dome and 3.0m (9.8ft) to keel
Gun Armament: One 100mm (3.9in) L/55 and two 20mm (0.79in) AA
Anti-submarine Armament: One Mk 54 sextuple 375mm (14.76in) rocket-launcher, and four tubes for 550mm (21.65in) L3 or 533mm (21in) L5 torpedoes
Propulsion: Two SEMT-Pielstick PCV diesels delivering 8,950kW (12,000bhp) to two shafts
Performance: Maximum speed 24kts; range 8,340km (5,185 mls) at 15kts, endurance 15 days
Complement: 5+29+45
Class
1. Argentina (3)
Name:

DRUMMOND Commissioned November 1978
2. France (15 + 2)

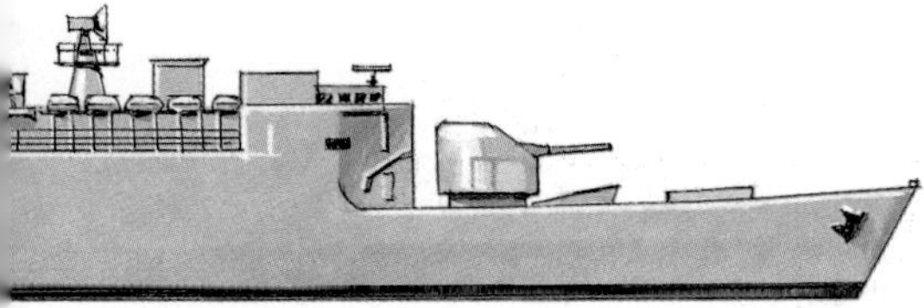

Name:
D'ESTIENNE D'ORVES Commissioned September 1976
Remarks: Officially rated as corvettes, the French A 69 class frigates are designed for coastal anti-submarine operations. However, 10 of the planned 18 ships have substantial anti-ship capability through the provision of Exocet surface-to-surface missile capability in all ships, although only those serving with the Mediterranean Fleet are actually fitted with the container-launchers. It is planned to upgrade all missile-carrying ships with MM.40 Exocets. The class has been found to possess excellent qualities of seaworthiness and economy, and this has promoted the use of the ships for a variety of alternative roles such as scouting, training and overseas deployments. Provision is made in the design for the accommodation of a troop detachment of one officer and 17 men.

ALMIRANTE PEREIRA DA SILVA CLASS FRIGATE

Country of Origin: Portugal
Displacement: 1,450 tons standard and 1,950 tons full load

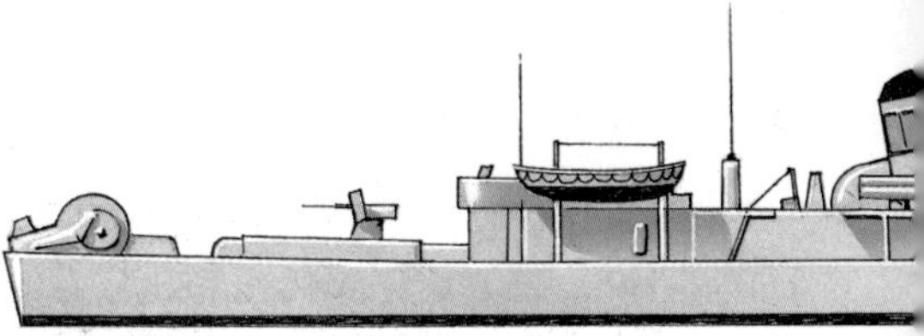

Dimensions: Length 95.9m (314.6ft); beam 11.26m (36.9ft); draught 5.3m (17.4ft)
Gun Armament: Four 76mm (3in) L/50 in two twin mountings
Anti-submarine Armament: Two triple Mk 32 tube mountings for 324mm (12.75in) Mk 44 A/S torpedoes, and two four-barrel Bofors 375mm (14.76in) rocket launchers
Propulsion: Two Foster-Wheeler boilers supplying steam to De Laval geared turbines delivering 14,915kW (20,000shp) to one shaft
Performance: Maximum speed 27kts; range 5,975km (3,715mls) at 15kts
Complement: 12+154
Class

1. Portugal (3)

ALMIRANTE PEREIRA DA SILVA Commissioned December 1966

Remarks: The Portuguese Almirante Pereira da Silva class is based on the US Dealey class design, fairly extensively modified to suit Portuguese operational requirements. The class has been in service for 20 years, and is ripe for replacement as maintenance is becoming difficult and the combined weapon/sensor fit is no longer adequate for Portugal's responsibilities within the NATO Atlantic requirement. Financial considerations at first dictated that the class be upgraded rather than replaced, but the emergence of NATO funding has permitted Portugal to undertake a new class of three modern frigates.

ALPINO CLASS
FRIGATE

Country of Origin: Italy
Displacement: 2,000 tons standard and 2,700 tons full load

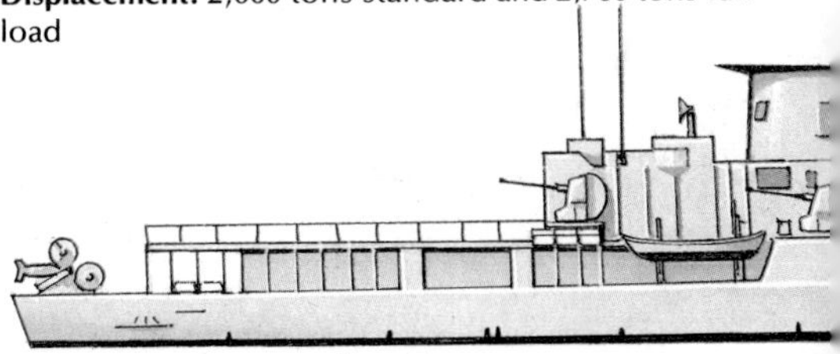

Dimensions: Length 113.3m (371.7ft); beam 13.3m (43.6ft); draught 3.9m (12.7ft)
Gun Armament: Six 76mm (3in) L/62 DP in six single mountings
Anti-submarine Armament: Two triple Mk 32 tube mountings for 324mm (12.75in) Mk 44 A/S torpedoes, one Mk 113 semi-automatic depth-charge thrower, and helicopter torpedoes
Aircraft: Two Agusta-Bell AB.212ASW helicopters
Electronics: One RCA SPS-12 surface-search radar, one SMA SPQ 2 search and navigation radar, three Orion gun fire-control radars as part of the Elsag/Argo 'O' control system, one SQS-43 hull-mounted sonar, one Litton SQA-10 variable-depth sonar, and one SCLAR overhead chaff/decoy/flare rocket system
Propulsion: CODAG (COmbined Diesel And Gas tur-

bine) arrangement with four Tosi diesels delivering 12,525kW (16,800hp) and two Tosi-Metrovick gas turbines delivering 11,185kW (15,000hp) to two shafts
Performance: Maximum speed 28kts on diesels and

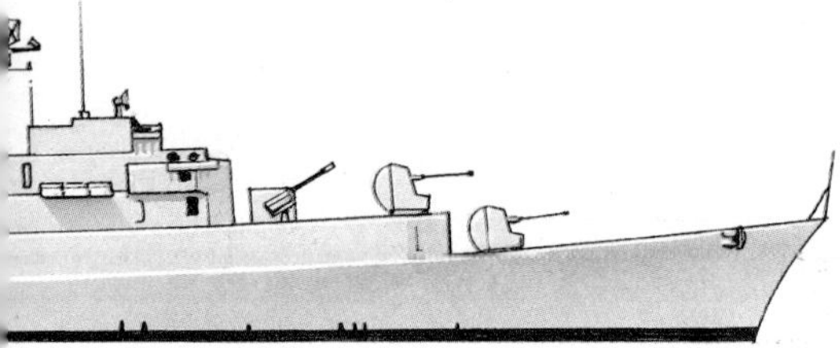

gas turbines, or 20kts on diesels; range 6,500km (4,040mls) at 18kts
Complement: 13+150
Class
1. Italy (2)
ALPINO Commissioned January 1968
CARABINIERE Commissioned April 1968
Remarks: The two Alpino class frigates are moderately effective escort ships, though their light anti-aircraft armament would relegate them more to convoy than to fleet work in the event of major hostilities. The combination of two Agusta-Bell AB.212ASW helicopters, variable-depth and hull-mounted sonars and an automatically-loaded and trainable depth-charge mortar provides fairly useful anti-submarine capability.

AMAZON CLASS (TYPE 21) GUIDED-MISSILE FRIGATE

Country of Origin: UK

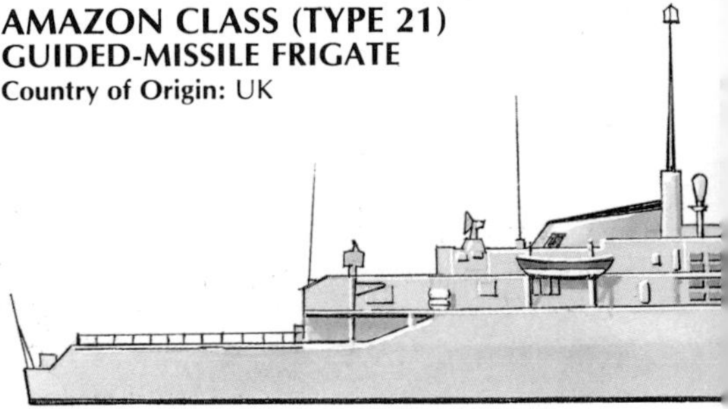

Displacement: 2,750 tons standard and 3,250+ tons full load

Dimensions: Length 117.0m (384ft); beam 12.7m (41.7ft); draught 5.9m (19.5ft)

Gun Armament: One 114mm (4.5in) L/55 in a single Mk 8 mounting, and two 20mm (0.79in) Oerlikon cannon in single mountings

Missile Armament: Four MM.38 Exocet surface-to-surface missiles in single launchers, and one quadruple launcher for Sea Cat surface-to-air missiles (to be replaced by GWS25 Sea Wolf possibly during second refit)

Anti-submarine Armament: Two triple tube mountings for 324mm (12.75in) Mk 46 A/S torpedoes, and helicopter-launched torpedoes

Aircraft: One Westland Wasp HAS.Mk 1 or Westland Lynx HAS.Mk2 helicopter

Propulsion: COGOG (COmbined Gas turbine Or Gas turbine) system, with two Rolls Royce Olympus TM3B gas turbines delivering 41,760kW (56,000bhp) or two Rolls-Royce Tyne RM1A gas turbines delivering 6,340kW (8,500shp) to two shafts
Performance: Maximum speed 30kts on Olympus turbines or 18kts on Tyne turbines; range 7,400km (4,600mls) at 17kts or 2,220km (1,380mls) at 30kts
Complement: 13+162, with a maximum of 192
Class
1. UK (6)
AMAZON Commissioned May 1974
Remarks: The Amazon or Type 21 class frigate was the first commercially-designed warship produced for the Royal Navy for many years, and was also the first British frigate class designed from the outset for gas turbine propulsion.

BALEARES CLASS (TYPE F70) GUIDED-MISSILE FRIGATE

Country of Origin: Spain
Displacement: 3,015 tons standard and 4,175 tons full load

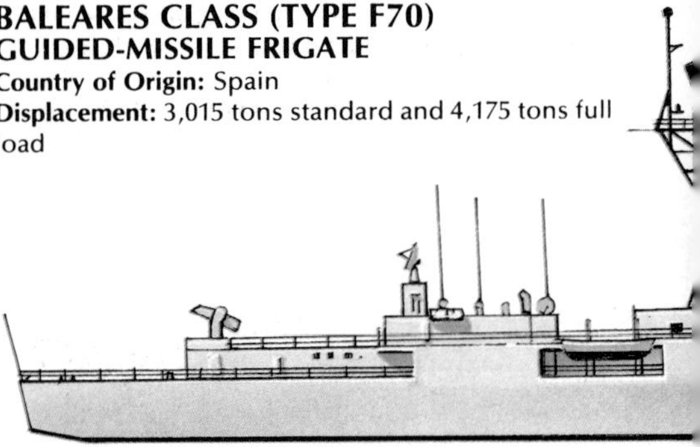

Dimensions: Length 133.6m (438ft); beam 14.3m (46.9ft); draught 4.7m (15.4ft)
Gun Armament: One 127mm (5in) L/54 in a Mk 42 mounting, and one Meroka AA and anti-missile point-defence system with two 20mm (0.79in) Oerlikon RTG six-barrel cannon
Missile Armament: One Mk 22 lightweight launcher for 16 RIM-24 Tartar or RIM-67 Standard ER surface-to-air missiles, and four or eight launchers for RGM-84A Harpoon surface-to-surface missiles
Anti-Submarine Armament: One RUR-5A ASROC launcher, four Mk 32 tubes for 324mm (12.75in) Mk 46 torpedoes, and two Mk 25 tubes for 484.5mm (19in) Mk 37 torpedoes; a maximum of 41 A/S torpedoes is carried

Propulsion: Two V2M-type boilers supplying steam to one set of Westinghouse geared turbines delivering 26,100kW (35,000shp) to one shaft
Performance: Maximum speed 28kts; range 8,350km

(5,190 mls) at 20kts
Complement: 15+241
Class
1. Spain (5)
Name:
BALEARES Commissioned September 1973
ANDALUCIA Commissioned May 1974
CATALUNA Commissioned January 1975
ASTURIAS Commissioned December 1975
EXTREMADURA Commissioned November 1976
Remarks: The Spanish navy's Baleares class frigate is modelled on the US Knox class, but has a different SAM launcher (with fewer missiles), Mk 25 anti-submarine torpedo tubes, no helicopter facilities, and revised sonar.

BREMEN CLASS (TYPE 122) GUIDED-MISSILE FRIGATE

Country of Origin: West Germany

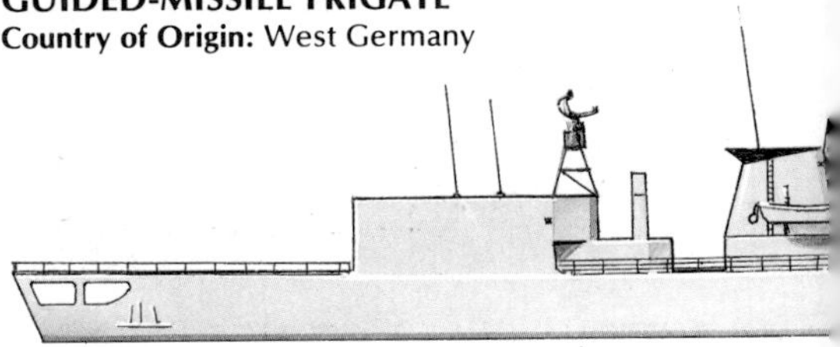

Displacement: 3,415 tons full load
Dimensions: Length 130.5m (428.1ft); beam 14.4 (48.5ft); draught 6.0m (19.7ft)
Gun Armament: Two quadruple launchers for eight RGM-84A Harpoon surface-to-surface missiles, one eight-cell launcher for RIM-7 Sea Sparrow surface-to-air missiles, two multiple launchers for FIM-92 Stinger surface-to-air missiles, and two launchers for RAM surface-to-air missiles
Anti-submarine Armament: Two twin Mk 32 tube mountings for 324mm (12.75in) Mk 44/46 A/S torpedoes, and helicopter-launched torpedoes
Aircraft: Two Westland Lynx helicopters
Propulsion: CODOG (COmbined Diesel Or Gas turbine) arrangement, with two General Electric LM 2500 gas turbines delivering 38,480kW (51,600hp) or two MTU 20V-956-TB92 diesels delivering 7,755kW (10,400hp) to two shafts

bines or 20kts on diesels; range 7,400km (4,600mls) at 18kts

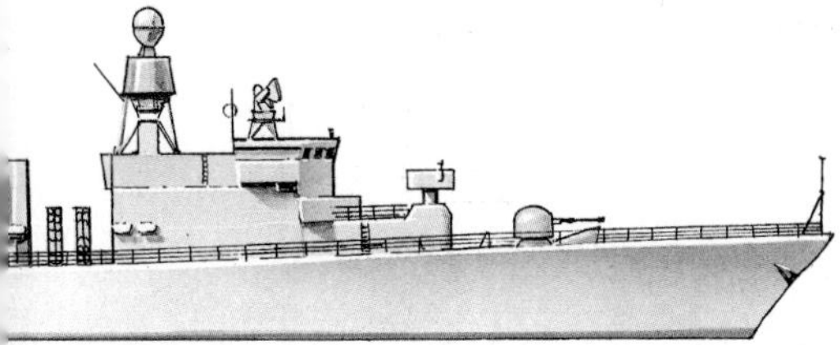

Complement: 203, with a maximum of 225 possible
Class
1. West Germany (4+2)
BREMEN Commissioned April 1982
Remarks: The Bremen or Type 122 guided-missile frigate has been developed from the Dutch Kortenaer class design to meet the West German navy's specific requirements for a frigate able to operate in high-threat areas as an effective anti-ship and anti-submarine system. The air and underwater threats are countered effectively by surface-to-air missile systems (Sea Sparrow, Stinger and RAM-ASDM) and Mk 32 tube mountings, while offensive capability is bestowed by the eight Harpoon anti-ship missiles and the two well-equipped Lynx anti-submarine helicopters. The 76mm (3in) OTO-Melara Compact gun and WM-25 fire-control system provide useful defence against fast attack craft.

BROADSWORD CLASS (TYPE 22)
GUIDED-MISSILE FRIGATE
Country of Origin: UK

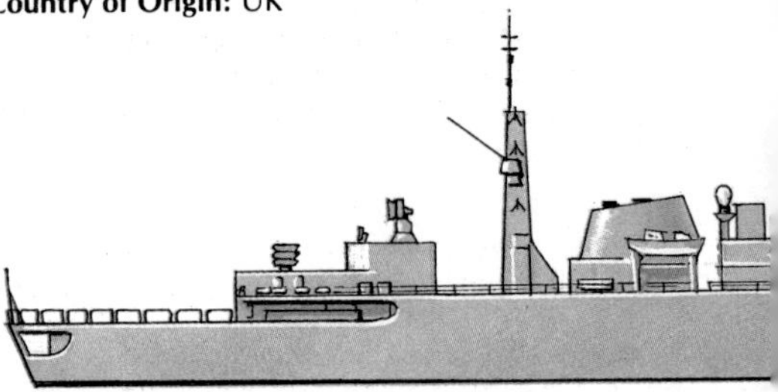

Displacement: 3,500 tons standard and 4,000 tons full load
Dimensions: Length 131.2m (430ft); beam 14.8m (48.5ft); draught 6.0m (19.9ft)
Gun Armament: Two 40mm (1.58in) Bofors L/60 AA in single mountings
Missile Armament: Four container-launchers for single MM38 Exocet surface-to-surface missiles, and two sextuple launchers for Sea Wolf surface-to-air missiles
Anti-submarine Armament: Helicopter-launched torpedoes
Aircraft: Two Westland Lynx HAS Mk 2 helicopters
Propulsion: COGOG (COmbined Gas turbine Or Gas turbine) arrangement, with two Rolls-Royce Olympus TM3B gas turbines delivering 41,760kW

(56,000shp) or two Rolls-Royce Tyne RM1A gas turbines delivering 6,340kW (8,500shp) to two shafts
Performance: Maximum speed 30kts (on Olympus engines) or 18kts (on Tyne engines); range 8,370km (5,200mls) at 18kts on Tyne engines
Complement: 18+205, with a maximum of 290 possible

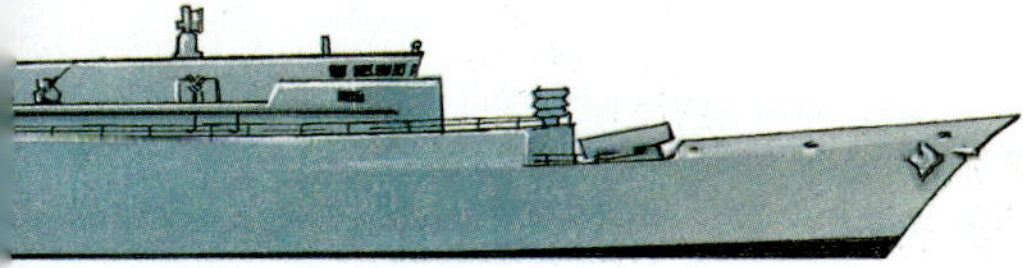

Class
1. UK (7+1+5)
BROADSWORD Commissioned May 1979
Remarks: Designed to succeed the widely admired Leander class, the Type 22 or Broadsword class guided-missile frigate has multi-role capability, though optimized, in present ships for anti-submarine operations. The first four units are Broadsword Batch 1 frigates, the following six ships being of the Broadsword Batch 2 type lengthened in the fore part of the ship and fitted with a more raked bow for better seakeeping qualities. The armament of Batch 1 and Batch 2 ships is similar, though from F 90 onwards the ships are each fitted with two triple STWS tube mountings for Mk 46 or Stingray anti-submarine torpedoes.

CHIKUGO CLASS
FRIGATE
Country of Origin: Japan

Displacement: 1,470-1,500 tons standard and 1,700 tons full load
Dimensions: Length 93.1m (305.5ft); beam 10.8m (35.5m); draught 3.5m (11.5ft)
Gun Armament: Two 3in (75mm) L/50 DP in one Mk 33 twin mounting, and two 40mm ((1.58in) Bofors AA in one Mk 1 twin mounting
Anti-submarine Armament: One octuple launcher for RUR-5A ASROC missiles, and two triple Type 68 tube mountings for 12.75in (324mm) Mk 44 torpedoes
Electronics: One OPS-14 air-search radar, one OPS-28 surface-search radar, one OPS-19 navigation radar, one Mk 33 radar used in conjunction with the Mk 1 gun fire-control system, one SQS-36 hull-mounted sonar and one SQS-35J variable-depth sonar
Propulsion: Four diesels delivering 11,930kW

(16,000shp) to two shafts
Performance: Maximum speed 25kts

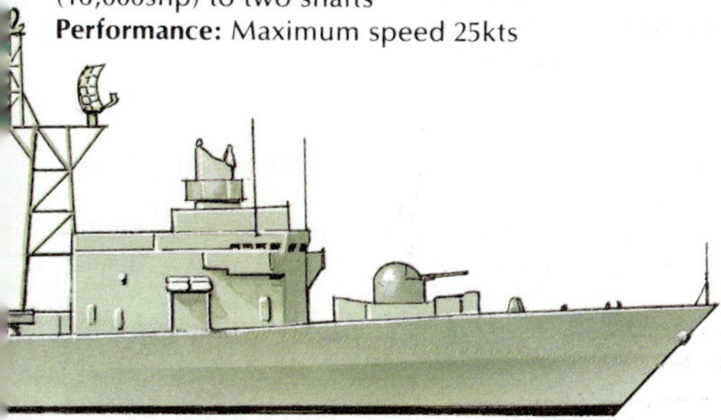

Complement: 165
Class
1. Japan (11)
CHIKUGO Commissioned July 1970
AYASE Commissioned May 1971
IWASE Commissioned December 1972
CHITOSE Commissioned August 1973
NIYODO Commissioned February 1974
TESHIO Commissioned January 1975
YOSHINO Commissioned February 1975
KUMANO Commissioned November 1975
NOSHIRO Commissioned August 1977
Remarks: Well-provided with sensors and electronic-warfare systems, the Chikugo class is the world's smallest warship type to carry the ASROC system.

COMMANDANT RIVIERE CLASS GUIDED-MISSILE FRIGATE

Country of Origin: France

Displacement: 1,750 tons standard and 2,250 tons full load

Dimensions: Length 103.0m (337.9ft); beam 11.5m (37.7ft); draught 4.3m (14.1ft)

Gun Armament: Two 100mm (3.9in) L/55 DP in single mountings, and two 30mm (1.18in)AA

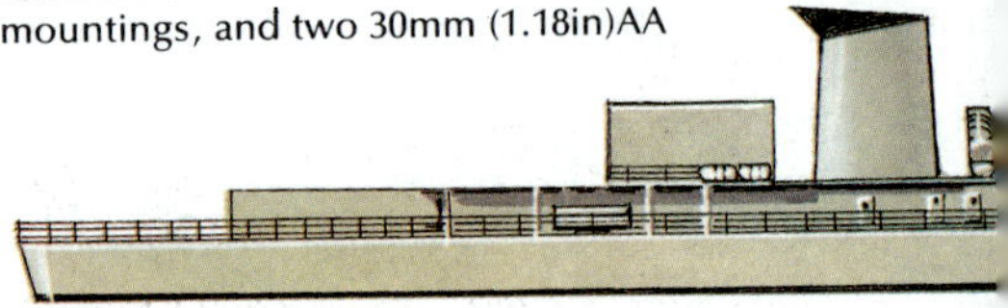

Missile Armament: Four container-launchers for MM38 Exocet surface-to-surface missiles

Anti-submarine Armament: Two triple 533mm (21in) tube mountings for L3 or K2 torpedoes, and one quadruple 305mm (12in) mortar

Aircraft: Provision for one light helicopter on a platform aft

Electronics: One Thomson-CSF DRBV 22A air-search radar, one Thomson-CSF DRBV 50 air/surface-search radar, one Thomson-CSF DRBC 32A gun-control radar, one Thomson-CSFDRBC 32C missile-control, one Thomson-CSF DRBN 32 navigation radar, one DUBA 3 hull-mounted sonar, one SQS-17 hull-mounted sonar and two CSEE Dagaie chaff launchers

Propulsion: Four SEMT-Pielstick diesels delivering

11,930kW (16,000bhp) to two shafts
Performance: Maximum speed 25kts; range 11,100km (6,700 mls) at 10-12kts
Complement: 10+61+96, and accommodation for a detachment of 80 troops who can be landed in two specially-carried LCPs

Class
1. France (9)
VICTOR SCHOELCHER Commissioned October 1962
COMMANDANT BORY Commissioned March 1964
AMIRAL CHARNER Commissioned December 1962
DOUDARD DE LAGRÉE Commissioned May 1963
BALNY Commissioned February 1970
COMMANDANT RIVIÉRE Commissioned December 1962
PROTET Commissioned May 1964
ENSEIGNE DE VAISSEAU HENRY Commissioned January 1965
Remarks: This well-balanced class of air-conditioned frigates is designed for worldwide operations, and as such each ship is fitted with command facilities.

DESCUBIERTA CLASS
GUIDED-MISSILE FRIGATE

Country of Origin: Spain

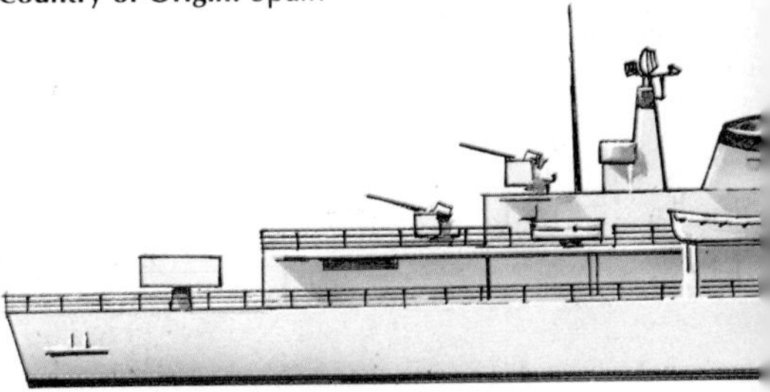

Displacement: 1,235 tons standard and 1,480 tons full load

Dimensions: Length 88.8m (291.3ft); beam 10.4m (34ft); draught 3.8m (12.5ft)

Gun Armament: One 76mm (3in) L/62 DP OTO Melara Compact, two 40mm (1.58in) Breda L/70 AA in single mountings, and one 20mm (0.79in) 12-barrel Meroka AA system

Missile Armament: Two quadruple launchers for eight RGM-84A Harpoon surface-to-surface missiles, and one Selenia Albatros octuple launcher sysem for 24 RIM-7 Sea Sparrow surface-to-air missiles

Anti-submarine Armament: Two triple Mk 32 tube mountings for 324mm (12.75in) Mk 46 torpedoes, and one 375mm (14.76in) Bofors twin rocket-launcher

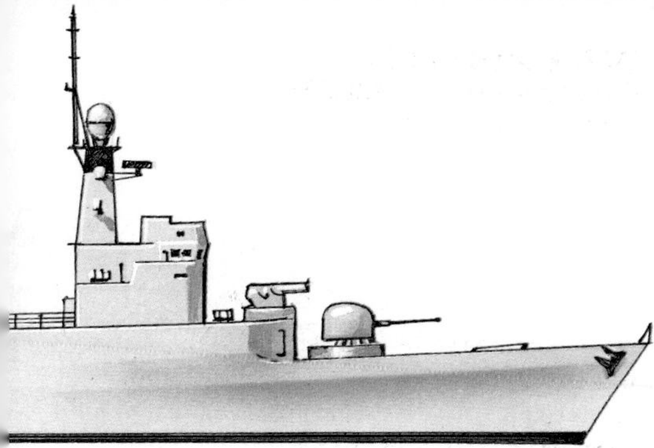

Propulsion: Four MTU/Bazan 16V956 TB91 diesels delivering 13,425kW (18,000bhp) to two shafts
Performance: Maximum speed 25kts; range 11,000km (6,700 mls) at 10-12kts
Complement: 116, plus provision for a marine detachment of 30
Class
1. Egypt (2)
2. Spain (6 F30 class)
DESCUBIERTA Commissioned November 1978
Remarks: Based on the Bazain-built Portuguese-proved Joaao Countintio class frigate, a 'Y' shaped funnel has replaced the original. An improved version is currently being designed with an increased displacement of 2,150 tons and facilities for a helicopter and hangar.

KONI CLASS
GUIDED-MISSILE FRIGATE

Country of Origin: USSR

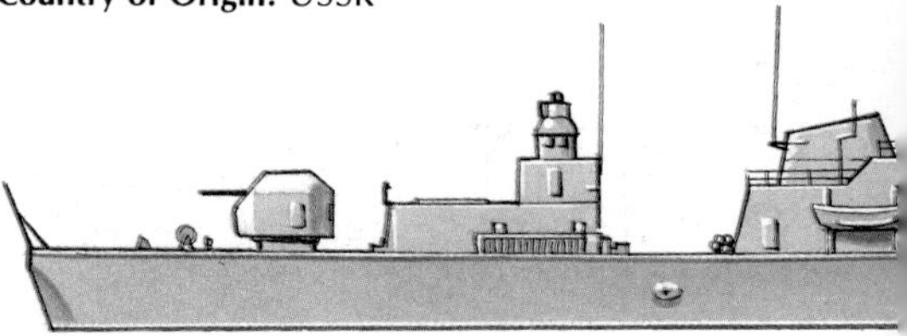

Displacement: 1,700 tons standard and 2,000 tons full load

Dimensions: Length 95.0m (311.6ft); beam 12.0m (39.3ft); draught 4.2m (13.7ft)

Gun Armament: Four 76mm (3in) L/60 DP in two twin mountings, and four 30mm (1.18in) AA in two twin mountings

Missile Armament: One twin launcher for SA-N-4 surface-to-air missiles

Anti-submarine Armament: Two RBU 6000 12-barrel rocket-launchers

Propulsion: CODAG (COmbined Diesel And Gas turbine) arrangement, with two diesels delivering 8,950kW (12,000shp) and one gas turbine delivering 13,425kW (18,000shp) to three shafts

Performance: Maximum speed 28kts on gas turbine and 22kts on diesels, range 3,700km (2,300 mls) at 14kts

Complement: 110

Class

1. Algeria (2)

MURAT REIS Commissioned December 1980

2. Cuba (1)

MARIEL Commissioned August 1981

3. East Germany (2)

ROSTOCK Commissioned July 1978

4. USSR (1)

DELFIN

5. Yugoslavia (1)

SPLIT Commissioned March 1980

Remarks: The small numbers of Koni class frigates produced in the USSR for export fall into two basically similar types, the Koni Type I and Koni Type II, the latter being distinguishable by the deckhouse filling the area between the after superstructure and funnel, presumably for air-conditioning equipment in vessels intended for hot-climate operations.

KORTENAER CLASS
GUIDED-MISSILE FRIGATE

Country of Origin: Netherlands
Displacement: 3,050 tons standard and 3,630 tons full load

Dimensions: Length 130.5m (428.1ft); beam 14.4m (47.2ft); draught 6.2m (20.3ft) to the screws
Gun Armament: One 76mm (3in) L/62 DP OTO-Melara Compact, and one 40mm (1.58in) Bofors AA (to be replaced by Hollandse Signaalapparaten close-in cannon system)
Missile Armament: Two quadruple container-launchers for eight RGM-84A Harpoon surface-to-surface missiles, and one Mk 29 launcher for RIM-7 Sea Sparrow surface-to-air missiles
Anti-submarine Armament: Two twin Mk 32 tube mountings for 324mm (12.75in) Mk 46 A/S torpedoes, and helicopter-launched weapons
Aircraft: Two Westland Lynx helicopters
Propulsion: COGOG (COmbined Gas turbine Or Gas turbine) arrangement with two Rolls-Royce Olympus TM3B gas turbines delivering 37,285kW (50,000shp) or two Rolls-Royce Tyne RMIC gas turbines delivering 5,965kW (8,000shp) to two shafts

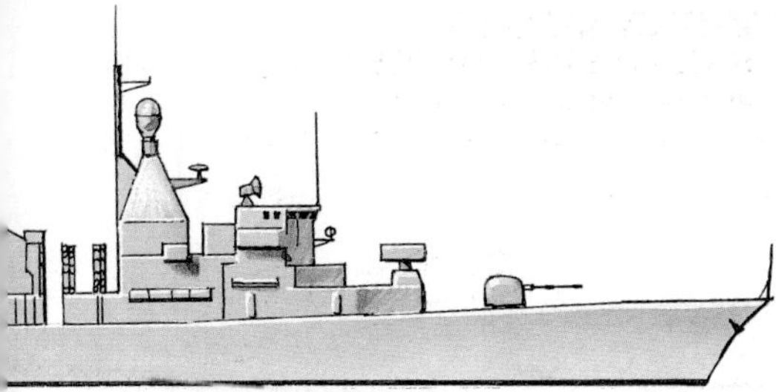

Performance: Maximum speed 30kts, range 8,700km (5,405mls) on Tynes at 16kts
Complement: 167
Class
1. Greece (2+1+2)
ELLI Commissioned October 1981
2. The Netherlands (10)
KORTENAER Commissioned October 1978
Remarks: The Kortenaer class is an excellent and versatile anti-submarine frigate design with useful anti-ship capability. The after 76mm (3in) gun is being removed in Dutch ships as the Goalkeeper CIWS mounting becomes available, and habitability is promoted by the extensive use of automation, which has reduced the crew from the 200 originally intended. The Greek ships (whose locally-built sisters appear to be in a construction limbo) have AB.212ASW helicopters, and a fire-control suite reduced to one WM-25 and one STIR equipment.

LEANDER (BATCH 3)
or BROAD-BEAM LEANDER CLASS FRIGATE

Country of Origin: UK

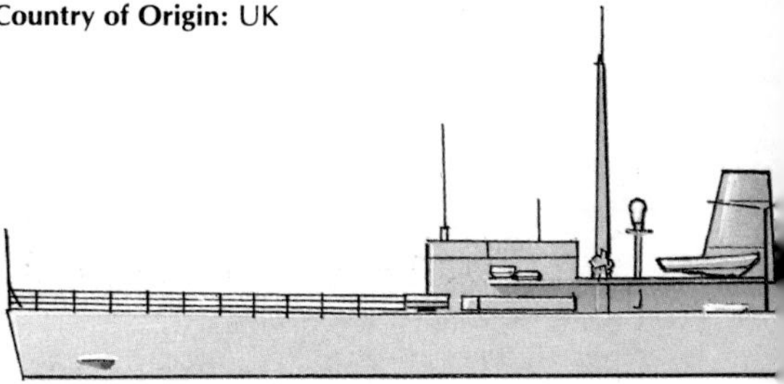

Displacement: 2,500 tons standard and 2,960 tons full load
Dimensions: Length 113.4m (372ft); beam 131.1m (43ft); draught 5.5m (18ft) to screws
Gun Armament: Two 114mm (4.5in) L/45 DP in one Mk 6 twin mounting and two 20mm (0.79in) AA in single mountings
Missile Armament: One quadruple launcher for Sea Cat surface-to-air missiles, and one sextuple launcher for Sea Wolf surface-to-air missiles
Anti-submarine Armament: One Limbo three-barrel mortar
Aircraft: One Westland Lynx HAS. Mk2 helicopter
Propulsion: Two Babcock & Wilcox boilers supplying steam to two sets of White-English Electric double-reduction geared turbines delivering 22,370kW

(30,000shp) to two shafts
Performance: Maximum speed 28kts; range 7,400km (4,600 mls) at 15kts

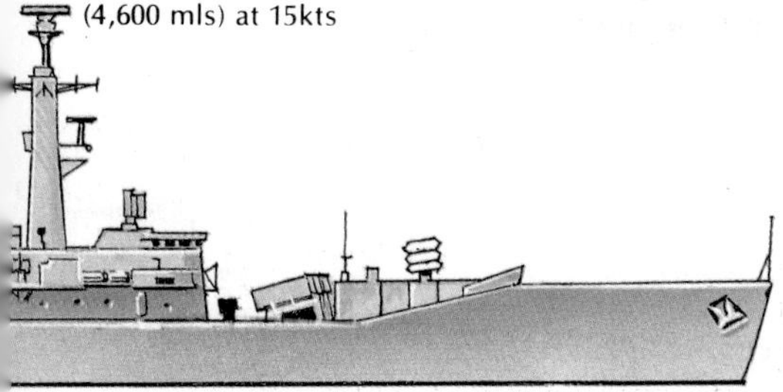

Complement: 19+241
Class:
1. Chile (2)
CONDELL Commissioned December 1973
2. India (6)
HIMGIRI Commissioned November 1974
3. New Zealand (2)
CANTERBURY Commissioned October 1971
4. UK (9)
ACHILLES
Remarks: The Leander class has been one of the most successful Western frigate designs since World War II, being built extensively in the UK and India, and forming the basis for the Dutch Van Speijk and Indian Godavari classes. The design was a development of the Type 12 class.

LUPO CLASS
GUIDED-MISSILE FRIGATE

Country of Origin: Italy
Displacement: 2,210 tons standard and 2,500 tons full load
Dimensions: Length 113.2m (371.3ft); beam 11.3m (37.1ft); draught 3.7 (12.1ft)

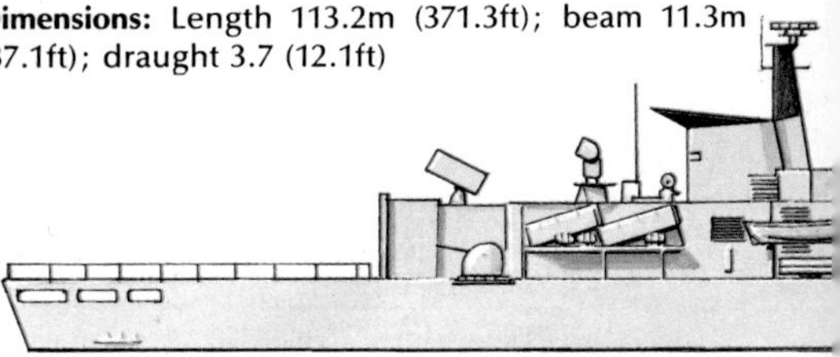

Gun Armament: One 127mm (5in) L/54 DP OTO-Melara Compact, and four 40mm Breda AA in two twin mountings
Missile Armament: Eight container-launchers for eight Otomat Mk 2 surface-to-surface missiles, and one octuple launcher for RIM-7 Sea Sparrow surface-to-air missiles
Anti-submarine Armament: Two triple Mk 32 mountings for 324mm (12.75in) Mk 44/46 A/S torpedoes, and helicopter-launched weapons
Aircraft: One helicopter in a hangar aft
Propulsion: CODOG (COmbined Diesel Or Gas turbine) arrangement, with two General Motors diesels delivering 5,815kW (7,800hp) or two Fiat/General Electric LM-2500 gas turbines delivering 37,285kW

(50,000hp) to two shafts
Performance: Maximum speed 35kts on gas turbines

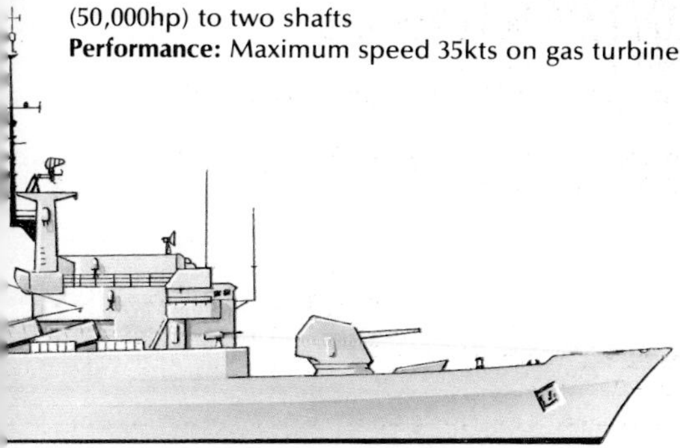

or 21kts on diesels; range 8,000km (4,970mls) at 16kts on diesels
Complement: 16+169
Class
1. Iraq (0+4)
2. Italy (4)
LUPO Commissioned September 1977
3. Peru (2+2)
MELITON CARVAJAL Commissioned February 1979
4. Venezuela (6)
MARISCAL SUCRE Commissioned May 1980
Remarks: The Lupo class has not proved altogether successful in Italian service despite its formidable armament.

MACKENZIE CLASS FRIGATE

Country of Origin: Canada

Displacement: 2,380 tons standard and 2,880 tons full load

Dimensions: Length 111.6m (366ft); beam 12.8m (42ft); draught 4.1m (13.5ft)

Gun Armament: Four 76mm (3in) L/50 DP in one Mk 6 and one Mk 33 twin mountings

Anti-submarine Armament: Two Limbo Mk 10 three-barrel mortars, and side-launchers for Mk 43 A/S torpedoes

Electronics: One Raytheon/Sylvania SPS-10 surface-search radar, one RCA SPS-12 air-search radar, one Bell SPG-48 gun-control radar used in conjunction with the Mk69 gun fire-control system, one Canadian Westinghouse SQS-501 bottom classification sonar, one Canadian Westinghouse SQS-502 mortar-control sonar, one Canadian Westinghouse SQS-503 search sonar, and one SQS-10 hull-mounted sonar

Propulsion: Two Babcock & Wilcox boilers supplying steam to two sets of English Electric geared turbines

delivering 22,370kW (30,000shp) to two shafts
Performance: Maximum speed 28kts; range 5,070km (3,150mls) at 14kts
Complement: 11+199
Class
1. Canada (4)
MACKENZIE Commissioned October 1962
YUKON Commissioned May 1963
Remarks: The four Mackenzie class frigates are officially classified as destroyers, and between 1982 and 1985 have been undergoing the Destroyer Life Extension Program refit to maintain them in service up to 1993. The design is unusual in that it uses two different 76mm (3in) gun mountings, and also because of the large stern accommodating a well for two Mk 10 Limbo mortars, the primary short-range anti-submarine armament.

MAESTRALE CLASS
GUIDED-MISSILE FRIGATE

Country of Origin: Italy

Displacement: 2,500 tons standard and 3,040 tons full load

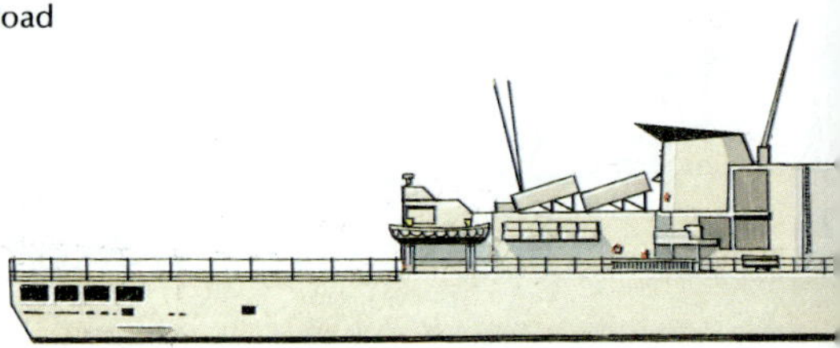

Dimensions: Length 122.7m (405ft); beam 12.9m (42.5ft); draught 8.4m (27.4ft) to screws

Gun Armament: One 127mm (5in) L/54 DP OTO-Melara Compact, and four 40mm (1.58in) Breda L/70 AA in two twin mountings

Missile Armament: Four container-launchers for four Otomat Mk 2 surface-to-surface missiles, and one quadruple launcher for Aspide surface-to-air missiles

Torpedo Armament: Two 533mm (21in) tubes for A 184 wire-guided anti-ship and anti-submarine torpedoes

Anti-submarine Armament: Two triple Mk 32 tube mountings for 324mm (12.75in) Mk 44/46 A/S torpedoes, and helicopter-launched weapons

Aircraft: Two Agusta-Bell AB.212ASW helicopters

Propulsion: CODOG (COmbined Diesel Or Gas turbine) arrangement, with two General Motors 230 die-

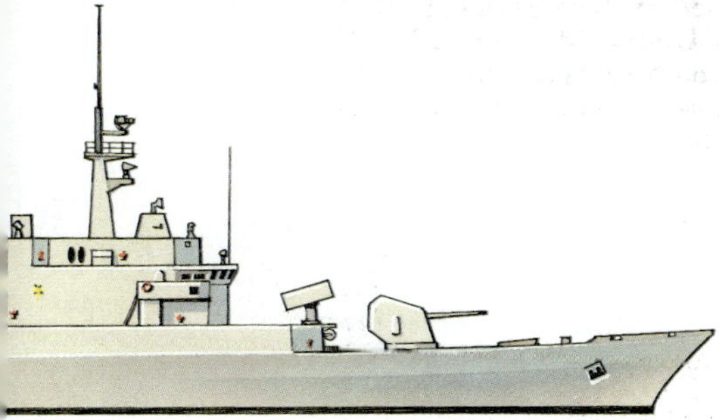

sels delivering 8,200kW (11,000hp) or two Fiat/General Electric LM 2500 gas turbines delivering 37,285kW (50,000shp) to two shafts
Performance: Maximum speed 32kts on gas turbines and 21kts on diesels; range 11,125km (6,915mls) at 16kts
Complement: 24+208
Class
1. Italy (6+2)
MAESTRALE Commissioned February 1982
Remarks: Developed from the Lupo design, the Maestrale class is an excellent fleet anti-submarine frigate, her extra size in comparison with the Lupo class permitting more comfortable accommodation, better seakeeping, a fixed hangar, and variable-depth sonar.

OLIVER HAZARD PERRY or FFG 7 CLASS
GUIDED-MISSILE FRIGATE
Country of Origin: USA

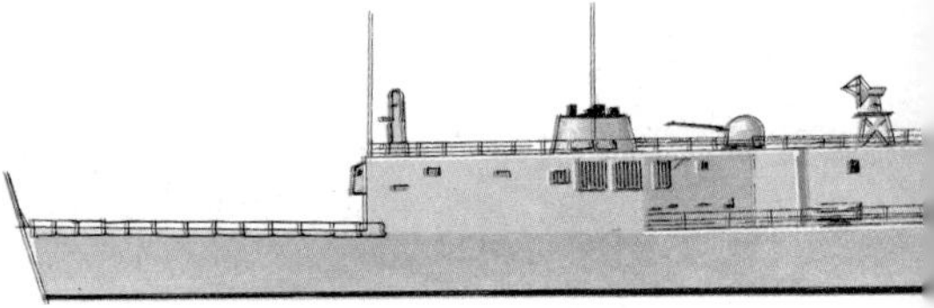

Displacement: 3,605 tons full load
Dimensions: Length 135.6m (445ft); beam 13.7m (45ft); draught 5.7m (24.5ft) to sonar dome
Gun Armament: One 76mm (3in) L/62 DP OTO-Melara Compact in a Mk 75 single mounting, and one 20mm (0.79in) Phalanx Mk 15 close-in weapon system mounting
Missile Armament: One Mk 13 single launcher for 40 RGM-84A Harpoon surface-to-surface and RIM-66 Standard-MR surface-to-air missiles
Anti-submarine Armament: Two triple Mk 32 tube mountings for 324mm (12.75in) Mk 46 A/S torpedoes, and helicopter-launched weapons
Aircraft: Two Kaman SH-2F Seasprite helicopters
Propulsion: Two General Electric LM 2500 gas turbines delivering 30,575kW (41,000shp) to one shaft
Performance: Maximum speed 29kts; range 8,370km (5,200mls) at 20kts

Complement: 11+153, and an air unit strength of 46

Class:

1. **Australia** (3+1+6)
2. **Spain** (0+3+2)
3. **USA** (35+10+5)

OLIVER HAZARD PERRY Commissioned December 1977

Remarks: The Oliver Hazard Perry class is one of the most significant frigate classes in the world, and is a balanced design for anti-submarine, anti-ship and anti-aircraft deployment. The planned 10 Australian ships are in all essential respects similar to the US class, while the much-delayed Spanish ships will have SQS-56 towed-array sonar and the Meroka close-in weapon system. In American ships the SQR-19 towed-array sonar is fitted only from FFG 36 onwards, the 2.4m (8ft) longer stern also permitting an improved helicopter haul-down system.

PEDER SKRAM CLASS
GUIDED-MISSILE FRIGATE

Country of Origin: Denmark
Displacement: 2,030 tons standard and 2,720 tons full load

Dimensions: Length 112.6m (396ft); beam 12.0m (39.5ft); draught 3.6m (11.8ft)
Gun Armament: Two 127mm (5in) L/38 DP in one Mk 38 twin mounting, and four 40mm (1.58in) Bofors L/60 AA in single mountings
Missile Armament: Two quadruple container-launchers for eight RGM-84A Harpoon surface-to-surface missiles, and one quadruple launcher for 16 RIM-7 Sea Sparrow surface-to-air missiles
Torpedo Armament: Two twin 533mm (21in) tube mountings
Anti-submarine Armament: Two depth-charge racks, and A/S torpedoes launched from the tubes above
Electronics: Two CWS-3 combined air-and surface-search radars, one NWS-1 tactical radar, one NWS-2 navigation radar, three CGS-1 fire-control radars, and one Plessey PMS 26 lightweight hull-mounted search

Propulsion: CODOG (COmbined Diesel Or Gas turbine) arrangement, with two General Motors 16-567D diesels delivering 3,580kW (4,800hp) and two Pratt & Whitney PWA GG4A-3 gas turbines delivering 32,810kW (44,000hp) to two shafts
Performance: Maximum speed 32.5kts on gas turbines or 16.5kts on diesels
Complement: 115
Class
1. Denmark (2)
PEDER SKRAM Commmissioned May 1966
HERLUF TROLLE Commissioned April 1967
Remarks: Despite their small size, the two Peder Skram class frigates are well-balanced anti-ship and anti-submarine vessels admirably suited to the operational requirements of the Royal Danish Navy.

RESTIGOUCHE (IMPROVED) CLASS FRIGATE

Country of Origin: Canada
Displacement: 2,390 tons standard and 2,900 tons full load
Dimensions: Length 113.1m (371ft); beam 12.8m (42ft); draught 4.3m (14.1ft)
Gun Armament: Two 76mm (3in) L/70 in one Mk 6 twin mounting

Anti-submarine Armament: One Limbo Mk 10 three-barrel mortar, and one octuple launcher for ASROC missiles
Propulsion: Two Babcock & Wilcox boilers supplying steam to two sets of English Electric geared turbines delivering 22,370kW (30,000shp) to two shafts
Performance: Maximum speed 28kts; range 8,810km (5,475 mls) at 14kts
Complement: 13+201
Class
1. Canada (4)
GATINEAU Commissioned February 1959
RESTIGOUCHE Commissioned June 1958

KOOTENAY Commissioned March 1959
TERRA NOVA Commissioned June 1959
Remarks: Officially rated as destroyers, the Restigouche (Improved) class frigates were evolved in the late 1960s and early 1970s by the elimination of the after 76mm (3in) twin mount and one Limbo mortar to provide space for an ASROC launcher and variable-depth sonar. A lattice foremast for the carriage of additional radars also replaced the tower-and-pole type of the basic Restigouche class, whose three units are now in reserve. The Destroyer Life Extension Program to be completed by 1986 will add new radar for service up to 1994.

TYPE FS 1500 CLASS
GUIDED-MISSILE FRIGATE

Country of Origin: West Germany
Displacement: 1,500 tons standard and 1,800 tons full load

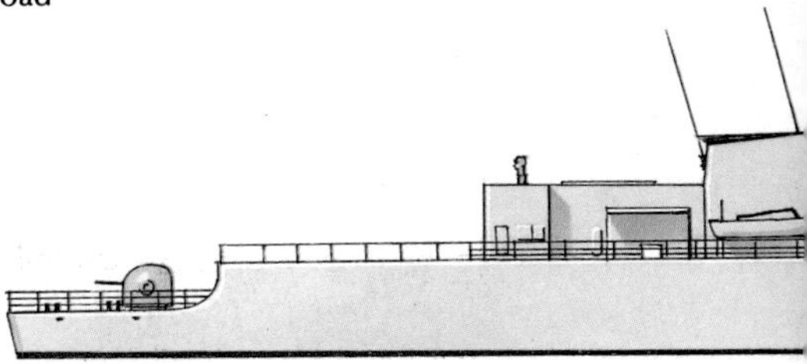

Dimensions: Length 90.0m (295.2ft); beam 11.3m (37.1ft); draught 3.4m (11.2ft)
Gun Armament: One 76mm (3in) L/62 DP OTO-Melara Compact, two 40mm (1.58in) Breda L/70 AA in a twin mounting, and two 30mm (1.18in) Oerlikon AA in a twin mounting
Missile Armament: Two quadruple container-launchers for eight MM.40 Exocet surface-to-surface missiles
Anti-submarine Armament: Two triple Mk 32 tube mountings for 324mm (12.75in) Mk 44/46 A/S torpedoes, and helicopter-launched weapons
Aircraft: One helicopter in a hangar aft
Propulsion: Four MTU 20V1163 TB62 diesels delivering 11,635kW (15,600hp) to two shafts

Fuel: Diesel oil
Performance: Maximum speed 26.5kts; range 9,250km (5,750mls) at 18kts
Complement: 90
Class
1. Colombia (3 + 1)
Three ships commissioned in 1983 having been laid down in 1981. A fourth is under construction
2. Malaysia (2+2)
2 on order with 2 more to follow
Remarks: The *Almirante Padilla* is the lead ship of the Colombian navy's force of four Type FS 1500 class guided-missile frigates which have been built to an impressive design by the Howaldtswerke Deutsche Wergt.

WIELINGEN CLASS
GUIDED-MISSILE FRIGATE

Country of Origin: Belgium
Displacement: 1,880 tons light and 2,285 tons full load
Dimensions: Length 106.4m (349ft); beam 12.3m (40.3ft); draught 5.6m (18.4ft)

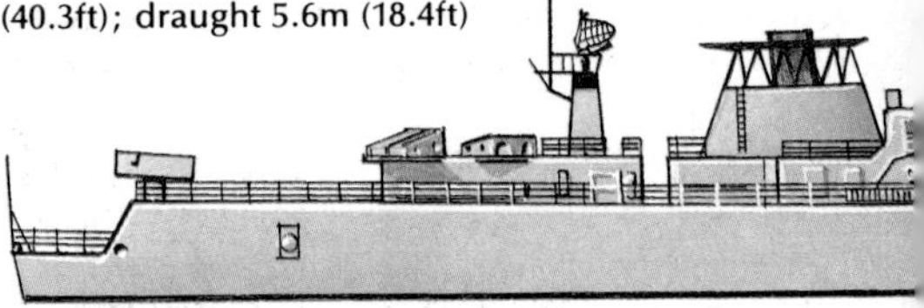

Gun Armament: One 100mm (3.9in) L/55 DP, and provision for one close-in weapon system mounting
Missile Armament: Four container-launchers for four MM.38 Exocet surface-to-surface missiles, and one octuple launcher for RIM-7 Sea Sparrow surface-to-air missiles
Torpedo Armament: Two launchers for L 5 torpedoes
Anti-submarine Armament: One 375mm (14.76in) Le Creuset-Loire six-barrel rocket launcher
Electronics: One Hollandse Signaalapparaten DA-05 combined air-and surface-search radar, one Raytheon TM 1645 navigation radar, one Hollandse Signaalapparaten WM-25 fire-control radar system, one CSEE Panda optical director, one Hollandse Signaalapparaten SEWACO IV data automation system, one ELCOS 1 radar-warning receiver, one Westinghouse SWS-505A hull-mounted medium-range

search and attack sonar, one Nixie torpedo decoy system, and two Mk 36 Super Chaffroc rapid-blooming overhead chaff launchers
Propulsion: CODOG (COmbined Diesel Or Gas turbine) arrangement, with two Cockerill CO-240 diesels delivering 4,475kW (6,000bhp) and one Rolls-Royce Olympus TM3B gas turbine delivering

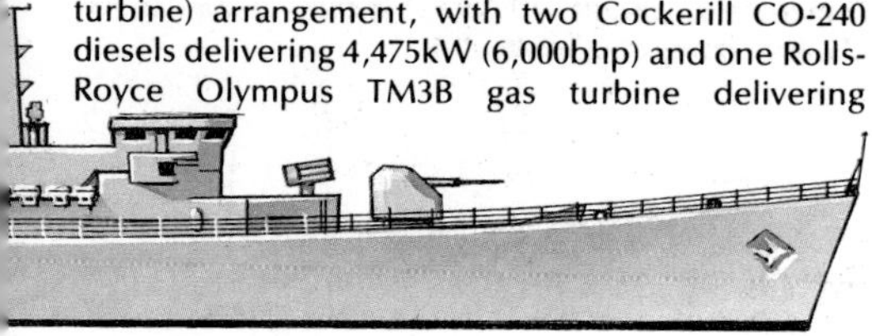

28,000bhp (20,880kW) to two shafts
Performance: Maximum speed 29kts on gas turbine and 20kts on diesels; range 8,350km (5,190 mls) at 18kts on diesels
Complement: 15+145
Class
1. Belgium (4)
WIELINGEN Commissioned January 1978
Remarks: Designed by the Belgian navy and built in Belgium, the Wielingen class marks a departure for the Belgian navy, which has hitherto been a coastal defence force. The type currently features two twin container-launchers for MM.38 Exocet anti-ship missiles on the after deckhouse, a position originally envisaged for the CIWS mounting when this latter is selected and fitted. The Signal/General Electric Goalkeeper system is likely to be chosen.

YUBARI CLASS
GUIDED-MISSILE FRIGATE

Country of Origin: Japan
Displacement: 1,400 tons
Dimensions: Length 91.0m (298.6ft); beam 10.8m (35.4ft); draught 3.5m (11.5ft)

Gun Armament: One 76mm (3in) L/62 DP OTO-Melara Compact
Missile Armament: Two quadruple container-launchers for eight RGM-84A Harpoon surface-to-surface missiles
Anti-submarine Armament: One 375mm (14.76in) Bofors four-barrel rocket-launcher, and two triple Type 68 tube mountings for 324mm (12.75in) Mk 44/46 A/S torpedoes
Electronics: Search/navigation radars, and sonar
Propulsion: CODOG (COmbined Diesel Or Gas turbine) arrangement, with one 6 DRV diesel delivering 3,505kW (4,700shp) and one Rolls-Royce Olympus TM3B gas turbine delivering 22,500 shp (16,780kW) to

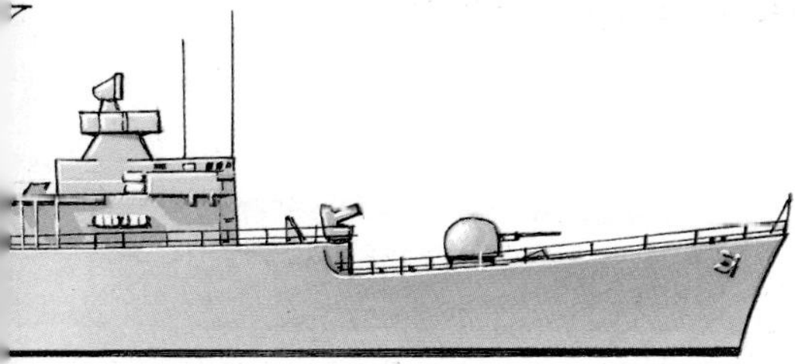

two shafts
Fuel: Diesel oil and kerosene
Performance: Maximum speed 26kts
Complement: 98
Class
1. Japan (2)
Name:
YUBARI Commissioned March 1983
YUBETSU Commissioned March 1984
Remarks: A multi-role frigate design evolved from the experimental Ishikari class, the Yubari class is currently planned at just two units, though three of a Yubari (Modified) class are planned as part of the 1983-87 five-year construction programme.

ASSAD CLASS
GUIDED-MISSILE CORVETTE

Country of Origin: Italy/Libya
Displacement: 670 tons full load
Dimensions: Length 61.7m (202.4ft); beam 9.3m (30.5ft); draught 2.2m (7.6ft)
Gun Armament: One 76mm (3in) L/62 DP OTO-Melara Compact, and two 35mm (1.38in) Oerlikon L/90 AA in one twin mounting

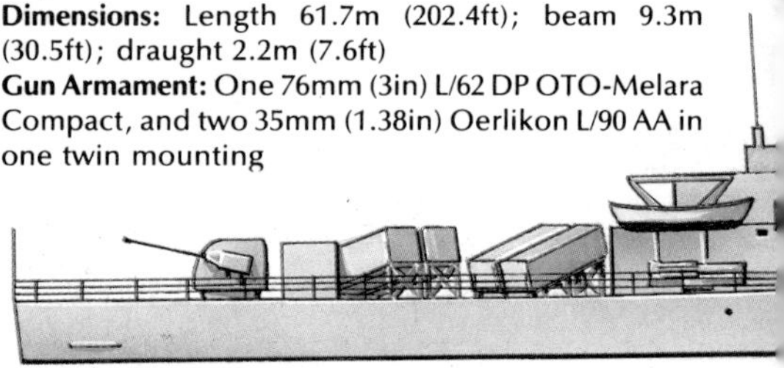

Missile Armament: Four container-launchers for four Otomat surface-to-surface missiles
Anti-submarine Armament: Two triple ILAS-3 tube mountings for 324mm (12.75in) A 244S A/S torpedoes
Mines: Up to 16
Propulsion: Four MTU 16V956 TB91 diesels delivering 13,425kW (18,000shp) to four shafts
Performance: Maximum speed 34kts; range 8,150km (5,065 mls) at 14kts
Complement: 58
Class
1. Ecuador (6)
ESMERALDAS Commissioned 1982

2. Iraq (1+5)
HUSSA EL HUSSAIR Commissioned 1984

3. Libya (4)
ASSAD EL TADJER Commissioned September 1979
Remarks: Originally designated the Wadi class when first ordered by Libya in 1974, this class was redesignated the Assad class during 1982-83 and has been ordered in useful numbers by two Arab countries (Iraq and Libya) and in a half-sister form by Ecuador. The Libyan standard is detailed above, and it is believed that the Iraqi ships will be essentially similar, though two of the six, namely the *Hussa el Hussair* (F 210) and the other unit building at CNR at Muggiano, have provision for a light helicopter in a telescopic hangar aft in place of four of the Otomat launchers.

DEIRDRE AND P22 CLASS
PATROL VESSEL

Country of Origin: Eire
Displacement: 1,020 tons

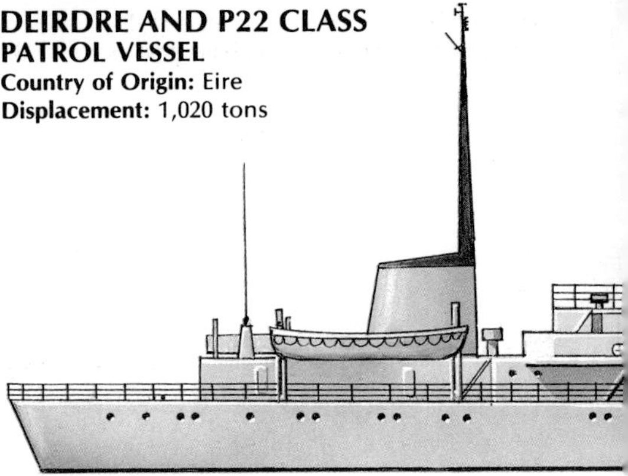

Dimensions: Length 65.2m (213.7ft); beam 10.5m (34.1ft); draught 4.4m (14ft)
Gun Armament: One 40mm (1.58in) Bofors AA, and two 20mm (0.79in) Oerlikon AA in single mountings
Propulsion: Two SEMT-Pielstick diesels delivering 3,580kW (4,800bhp) to one shaft
Performance: Maximum speed 18kts; range 12,515km (7,775mls) at 12kts
Complement: 5+41
Class
1. Eire (4)
DEIRDRE Commissioned May 1972
AISLING Commissioned May 1980

Remarks: The *Deirdre* was the first Eire-built ship for Irish naval service, and is slightly different from the

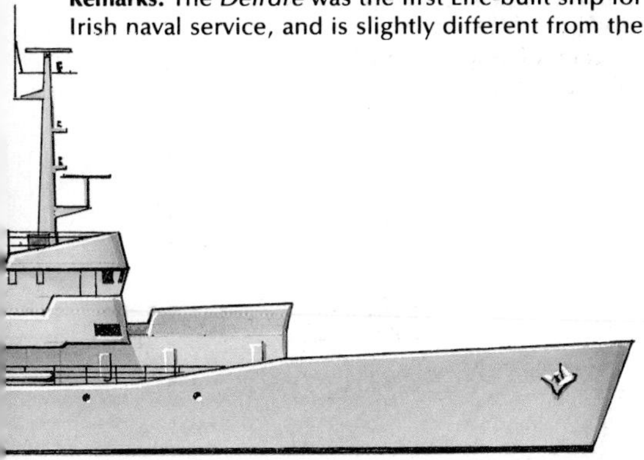

other three vessels, which are officially the P22 class to which the specification above applies. The *Deirdre* displaces 972 tons, has a length of 56.2m (184.3ft) between perpendiculars, and is capable of 18kts on the 3,130kW (4,200bhp) delivered by her two British Polar diesels. Her armament comprises a single 40mm (1.58in) gun which can operate effectively against small craft, and this is fitted forward of the bridge behind a three-sided spray shield. The provision of hull-mounted sonar gives the ships a submarine-detection capability, though helicopters would have to be called in for a 'kill'.

NANUCHKA I AND NANUCHKA III CLASS
GUIDED-MISSILE CORVETTES
Country of Origin: USSR

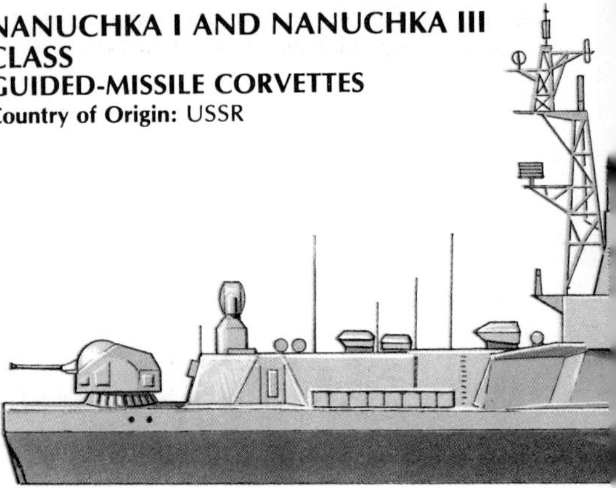

Displacement: 780 tons standard and 900 tons full load
Dimensions: Length 60.0m (196.8ft); beam 12.2m (40ft); draught 3.1m (10.2ft)
Gun Armament: Two 57mm (2.25in) L/70 AA in one twin mounting ('Nanuchka I') or one 76mm (3in) L/59 DP ('Nanuchka III'), and ('Nanuchka III' only) one 30mm (1.18in) AA in a 'Gatling' mounting
Missile Armament: Two triple container-launchers for six SS-N-9 surface-to-surface missiles, and one twin launcher for 18 SA-N-4 surface-to-air missiles
Propulsion: Three diesels delivering 17,895kW (24,000shp) to three shafts

Performance: Maximum speed 34kts; range 8,350km (5,190mls) at 15kts or 2,400km (1,490 mls) at 33kts
Complement: 70
Class
1. Algeria (2)

RAS HAMIDOU Commissioned July 1980
2. India (3+0+3)
VIJAY DURG Commissioned March 1977
3. Libya (2+2)
EAN MARA Commissioned October 1981
4. USSR (17 Nanuchka I and 5+1 Nanuchka III class)
Remarks: The Nanuchka classes are extremely powerful coastal vessels built in the Pacific and at Petrovsky from 1969. The Indian ships differ mainly in having four SS-N-2B anti-ship missiles in place of the longer-range SS-N-9s.

VOSPER THORNYCROFT MK 9 CLASS CORVETTE

Country of Origin: UK/Nigeria
Displacement: 850 tons full load

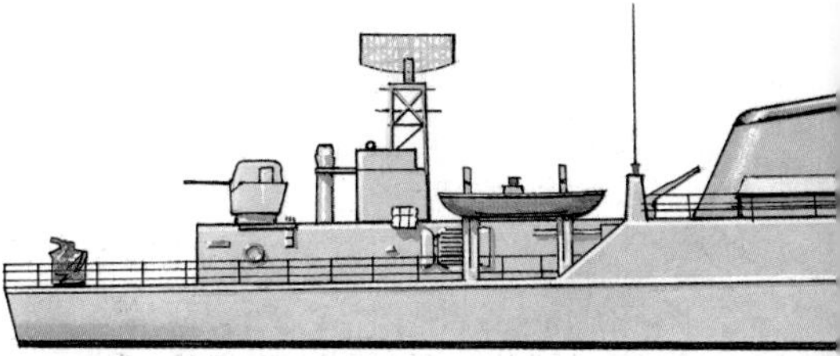

Dimensions: Length 69.0m (226ft); beam 9.6m (31.5ft); draught 3.0m (9.8ft)
Gun Armament: One 76mm (3in) L/62 DP OTO-Melara Compact, one 40mm (1.58in) Bofors L/70 AA, and two 20mm (0.79in) Oerlikon AA
Missile Armament: One triple launcher for 12 Sea Cat surface-to-air missiles
Anti-submarine Armament: One 375mm (14.76in) Bofors two-barrel rocket-launcher
Electronics: One Plessey AWS 2 surface-search radar, one Decca TM 1226 navigation radar, one Hollandse Signaalapparaten WM-24 fire-control radar system, and one Plessey PMS 26 hull-mounted sonar
Propulsion: Four MTU 20V956 TB92 diesels delivering

13.125kW (17,600shp) to two shafts
Performance: Maximum speed 27kts; range 4,105km (2,550 mls) at 14kts
Complement: 90
Class
1. Nigeria (2)
Name:
ERIN'MI Commissioned January 1980
ENYIMIRI Commissioned July 1980
Remarks: The two Mk9 craft were ordered from Vosper Thornycroft in April 1975 and were laid down in January and February 1977. During 1979 they both underwent minor refits including the heightening of their funnels.

CONSTITUCION CLASS
FAST ATTACK CRAFT (MISSILE/GUN)

Country of Origin: UK
Displacement: 170 tons

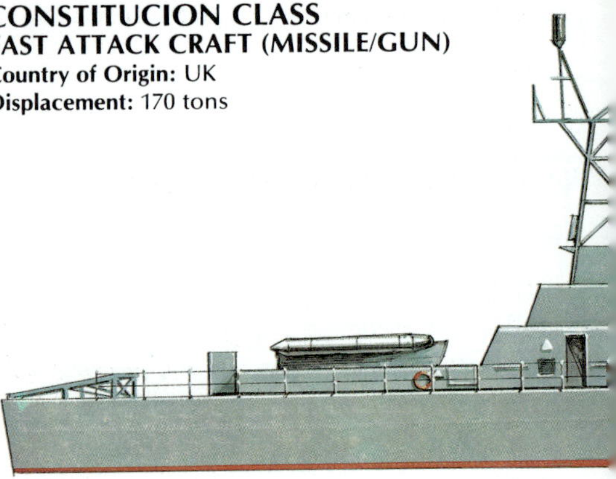

Dimensions: Length 36.9m (121ft); beam 7.1m (23.3ft); draught 1.8m (6ft)
Gun Armament: One 76mm (3in) L/62 DP OTO-Melara Compact and one 40mm (1.58) Breda L/70 AA
Missile Armament: Two container-launchers for two Otomat surface-to-surface missiles
Electronics: One SMA SPQ 2D surface-search radar, and one Selenia Orion RTN 10X fire-control radar used in conjunction with the ELSAG Argo NA 10 fire-control system
Propulsion: Two MTU diesels delivering 5,370kW (7,200hp) to two shafts

Performance: Maximum speed 31kts; range 2,500km (1,555 mls) at 16kts
Complement: 3+14
Class
1. Venezuela (6)
Name:
CONSTITUCION Commissioned August 1974

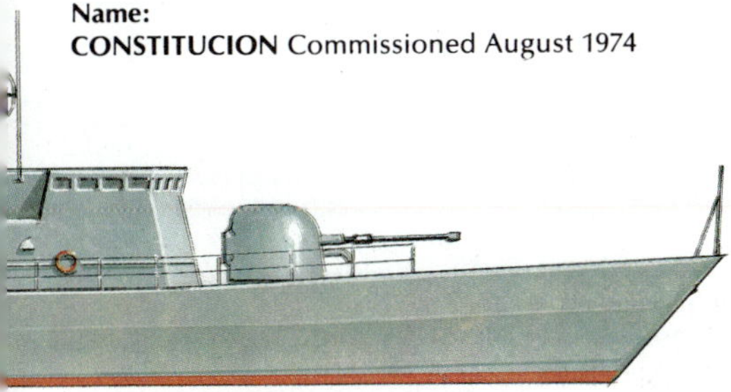

FEDERACION Commissioned March 1975
INDEPENDENCIA Commissioned September 1974
LIBERTAD Commissioned June 1975
PATRIA Commissioned January 1975
VICTORIA Commissioned September 1975
Remarks: Also known as the Vosper Thornycroft 121ft class the order was valued at 6 million when placed in 1972. The six craft were built in a fourteen month period from January 1973. *Federacion, Libertad* and *Victoria* were built as missile craft although it has since been reported that they have been removed from *Federacion*.

CORMORAN CLASS
FAST ATTACK CRAFT (MISSILE)
Country of Origin: Spain/Morocco

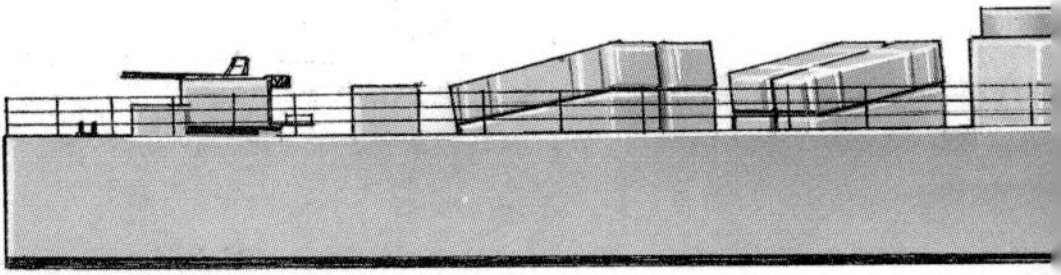

Displacement: 410 tons full load
Dimensions: Length 57.4m (188.3ft); beam 7.6m (24.9ft); draught 2.7in (8.9ft)
Gun Armament: One 76mm (3in) L/62 DP OTO-Melara Compact, and one 40mm (1.58in) Breda L/70 AA
Missile Armament: Four container-launchers for MM.40 Exocet surface-to-surface missiles
Electronics: One Hollandse Signaalapparaten ZW-06 search radar, one Hollandse Signaalapparaten Wm-20 fire-control radar, and one CSEE director system
Propulsion: Two MTU/Bazan 16V356 T1391 diesel delivering 6,000kW (8,045bhp) to two shafts
Performance: Maximum speed 36kts; range 4,650km (2,890 mls) at 15kts
Complement: 41

Class

1. Morocco (4)

Name:

EL KHATTABI Commissioned July 1981

COMMANDANT BOUTOUBA Commissioned November 1981

COMMANDANT EL HARTY Commissioned February 1982

COMMANDANT AZOUGGARGH Commissioned August 1982

Remarks: Built by the Farran Spanish Bazan Yards at Cadiz, the four craft were originally designated Lazaga class. Based at Casablanca, they form a vital part of the Moroccan navy which only has 23 active craft and less than 2000 personnel including marines.

LA COMBATTANTE III CLASS
FAST ATTACK CRAFT (MISSILE AND TORPEDO)

Country of Origin: France/Greece
Displacement: 360 tons standard and 425 tons full load

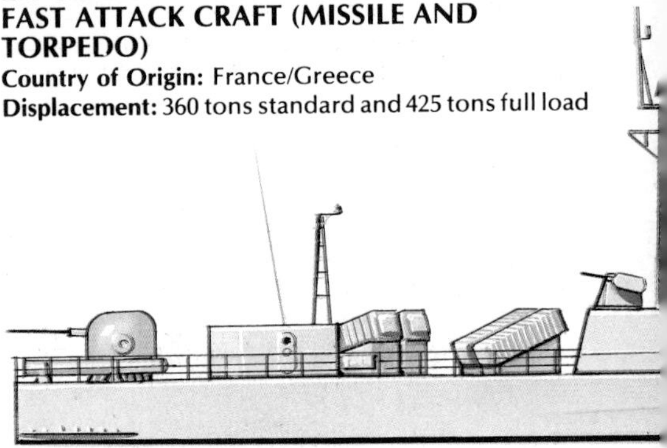

Dimensions: Length 56.0m (183.7ft); beam 7.9m (25.9ft); draught 2.5m (8.2ft)
Gun Armament: Two 76mm (3in) L/62 DP OTO-Melara Compact in two single mountings and four 30mm (1.18in) AA in two Emerlec twin mountings
Missile Armament: Four container-launchers for MM.38 Exocet surface-to-surface missiles
Torpedo Armament: Two 533mm (21in) tubes
Electronics: One Thomson-CSF Triton surface-search and navigation radar, one Thomson-CSF Vega II SSM fire-control system with Thomson-CSF Castor and Thomson-CSF Pollux tracking radars, and two CSEE Panda optical fire directors

Propulsion: Four MTU MD 20V538 TB91 diesels delivering 13,425kW (18,000bhp) to four shafts
Performance: Maximum speed 35.7kts; range 3,700km (2,300 mls) at 15kts or 1,300km (810mls) at 32.6kts

Complement: 5+37
Class
1. Greece (10)
Name:
ANTIPLOIARHOS LASKOS Commissioned April 1977
Remarks: The *Antiploiarhos Laskos* is lead craft of the 10-strong La Combattante III class of Greek navy fast-attack craft, the first four of which were ordered in September 1974, with *Antiploiarhos Laskos* being launched on 6 July 1976. She was followed by *Ploturhis Blessas,* launched 10 November 1976. The second group of six were ordered in 1978. The primary armament comprises four MM.38s.

LURSSEN FPB 57 CLASS
FAST ATTACK CRAFT (MISSILE)

Country of Origin: West Germany/Nigeria
Displacement: 410 tons full load
Dimensions: Length 58.1m (190.6ft); beam 7.6m (24.9ft); draught 2.7m (8.9ft)

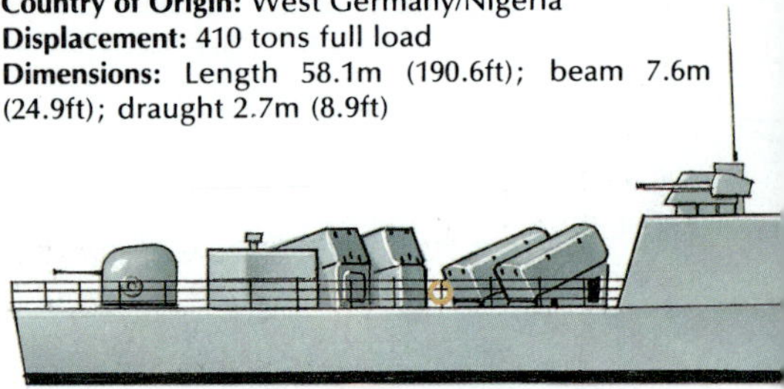

Gun Armament: One 76mm (3in) L/62 DP OTO-Melara Compact, and two 40mm (1.58in) Breda L/70 AA and two twin Emerlec 30mm (1.18in) mountings
Missile Armament: Four Otomat anti-ship missiles
Propulsion: Four MTU 16V956 TB92 diesels delivering 15,000kW (20,115shp) to two shafts
Performance: Maximum speed 42kts; range 2,400km (1,490 mls) at 30kts
Complement: 40
Class
1. **Kuwait** (2)
2. **Nigeria** (3)
EKPE Commissioned August 1981
3. **Singapore** (3)
4. **Turkey** (5)
DOGAN Commissioned June 1977

Remarks: The largest standard-hull type produced by Lurssen, perhaps the world's foremost designer of fast attack craft, the Lurssen FBP 57 serves with three navies as a missile craft. The specification applies to the Nigerian craft, but all three sub-types are similar in size and propulsion. The Kuwaiti duo have 13,400kW (17.970shp) for 36kts, and the armament – consisting of four MM.40 Exocet anti-ship missiles, one 76mm (3in) gun and one 40mm (1.58in) Breda twin mounting – has a Philips Elektronikindustrier 9LV 228 fire-control system. The Turkish craft differ in having eight RGM-84A Harpoon anti-ship missiles in two quadruple container-launchers, and a secondary gun armament of one 35mm (1.38in) twin mounting. Singapore's three craft are FAC(G)s with one 76mm (3in) and two 40mm (1.58in) guns.

LURSSEN TNC 45 CLASS
FAST ATTACK CRAFT (MISSILE)

Country of Origin: West Germany/UAE
Displacement: 230 tons
Dimensions: Length 45.0m (147.6ft); beam 7.0m (23ft); draught 2.3m (7.5ft)
Gun Armament: One 76mm (3in) L/62 DP OTO-Melara

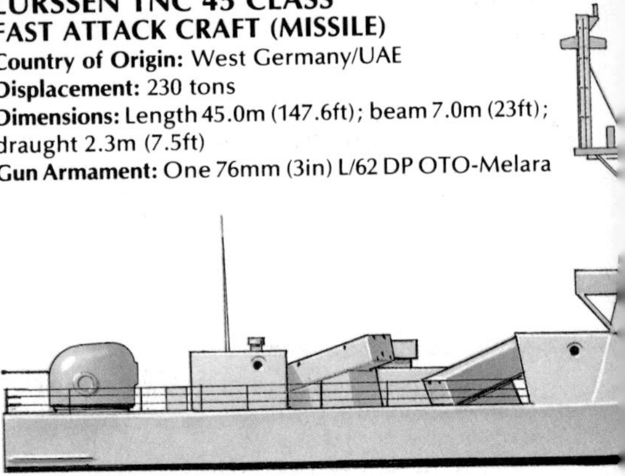

Compact, and two 40mm (1.58in) Breda AA in a twin mounting
Missile Armament: Four container-launchers for four MM.40 Exocet surface-to-surface missiles
Propulsion: Four MTU diesels delivering 10,740kW (14,400hp) to four shafts
Performance: Maximum speed 40kts
Complement: 40
Class
1. **Argentina** (2)
2. **Bahrain** (0+2)

3. **Indonesia** (2)
4. **Kuwait** (6)
5. **Singapore** (6)
6. **Thailand** (3)
7. **United Arab Emirates** (6)

Remarks: The Lurssen TNC 45 is one of the most successful FAC classes yet developed, the specifica-

tion above applying to the variant in service with the United Arab Emirates. Argentina's two craft were delivered as FAC(G/T)s with one 76mm (3in) and two 40mm (1.58in) guns and two 533mm (21in) torpedo tubes; they are currently being converted to carry Otomat anti-ship missiles. Bahrain's two craft carry four Exocets, one 75mm (3in) gun and one twin 40mm (1.58in) mounting apiece, and have WM-28 fire control; both craft were delivered from Lurssen by 1984. Indonesia's two craft are FAC(T)s, comparable to the West German Jaguar class, with four 533mm (21in) tubes.

OCTOBER CLASS
FAST ATTACK CRAFT (MISSILE)

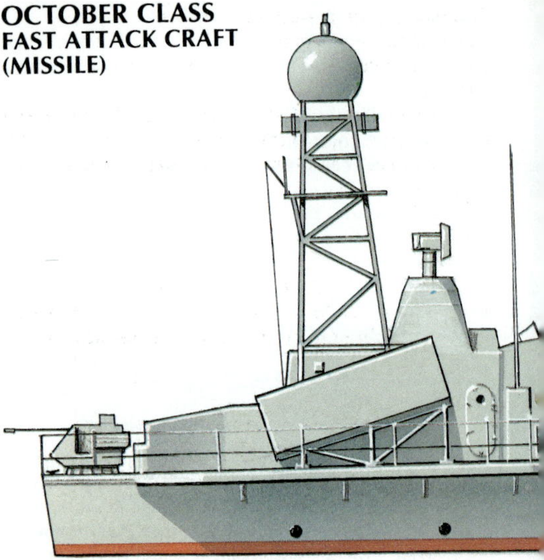

Country of Origin: Egypt
Displacement: 82 tons full load
Dimensions: Length 25.5m (84ft); beam 6.1m (20ft); draught 1.3m (5ft)
Gun Armament: Four 30mm (1.18in) AA in two twin mountings
Missile Armament: Two container-launchers for two

Otomat surface-to-surface missiles

Electronics: One Marconi S810 surface-search radar, and one Marconi/Sperry Sapphire missile-control radar

Propulsion: Four CRM 18D/S2 diesels delivering 4,025kW (5,400hp) to four shafts

Performance: Maximum speed 40kts; range 750km (465 mls) at 30kts

Complement: 20

Class

1. Egypt (6)

6 craft (nos 207, 208, 209, 210, 211 and 212)

Remarks: The October class FAC(M) is an interesting hybrid, combining the hull of the Soviet Komar class

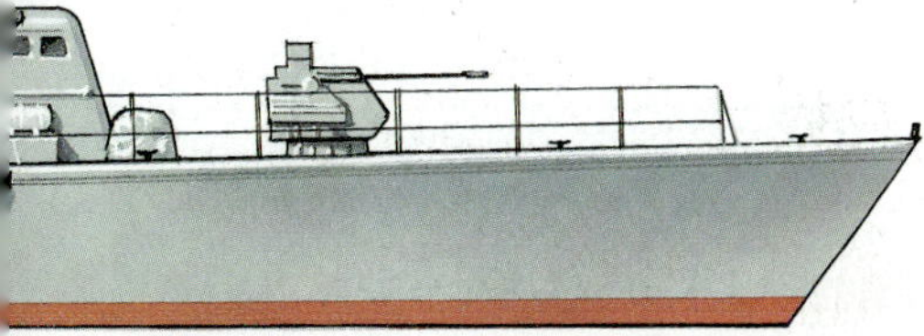

with armament and electronics of Western European manufacture. The armament is relatively light, this being necessitated by the small size and displacement of the class, but in conjunction with high speed and manoeuvrability this still gives the Egyptian craft good capability for coastal operations. The craft were built in Alexandria between 1975 and 1976.

PEGASUS CLASS
FAST ATTACK HYDROFOIL (MISSILE)

Country of Origin: USA
Displacement: 240 tons full load
Dimensions: Length 40.5m (132.9ft) with foils extended and 44.3m (145.3ft) with foils retracted; beam 14.5m (47.5ft) with foils extended and 8.6m (28.2ft) with foils retracted; draught 7.1m (23.3ft) with foils extended and 2.3m (7.5ft) with foils retracted
Gun Armament: One 76mm (3in) L/62 DP OTO-Melara Compact

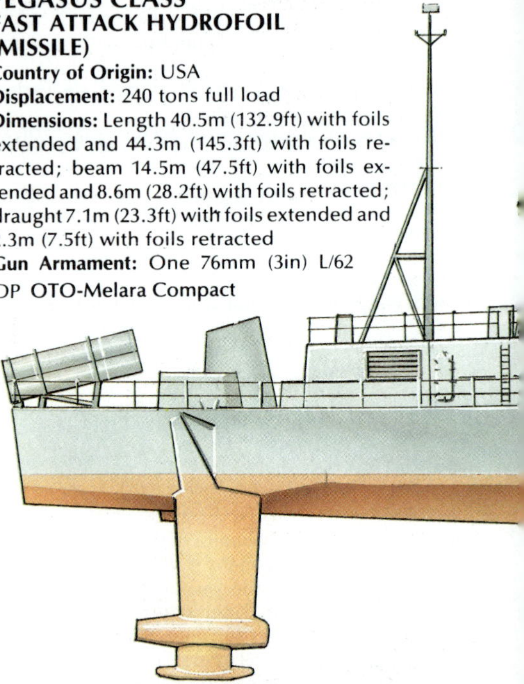

Missile Armament: Four RGM-84A Harpoon surface-to-surface missiles
Propulsion: Two MTU 8V331 TC81 diesels delivering 1,195kW (1,600bhp) to two water-jets for hullborne operation and one General Electric LM 2500 gas turbine delivering 13,425kW (18,000shp) to Aerojet waterjets for foilborne operation
Performance: Maximum speed 48kts foilborne and 12kts hullborne; range 3,140km (1,950 mls) at 9kts and 1,285km (800 mls) at 40kts

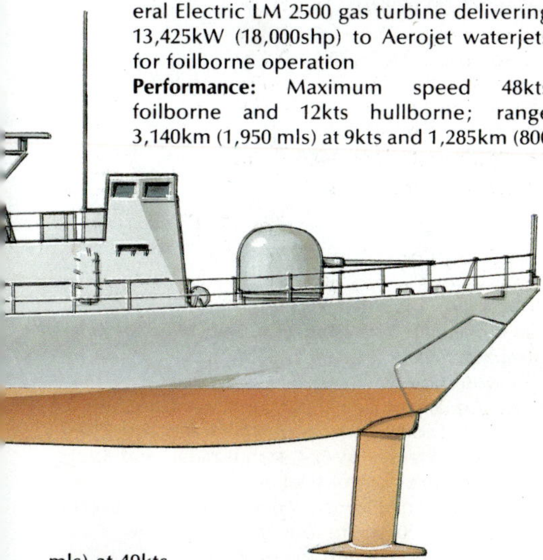

Complement: 4+17
PEGASUS Commissioned July 1977

PROVINCE CLASS
FAST ATTACK CRAFT (MISSILE)

Country of Origin: UK/Oman
Displacement: About 420 tons full load
Dimensions: Length 56.7m (186ft); beam 8.2m (26.9ft); draught 2.7m (8.9ft)

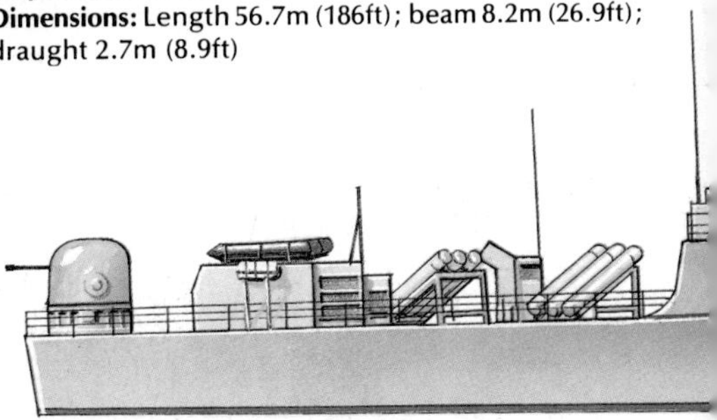

Gun Armament: One 76mm (3in) L/62 DP OTO-Melara Compact, and two 40mm (1.58in) Breda L/70 AA in a twin mounting
Missile Armament: Two triple container-launchers for six MM.40 Exocet surface-to-surface missiles
Electronics: Plessey AWS 4 search radar, and Sperry Sea Archer fire-control system
Propulsion: Four Paxman Valenta diesels delivering 13,570kW (18,200hp) to four shafts
Performance: Maximum speed 40kts
Complement: 59, including 19 passengers/trainees
Class

1. Oman (2+1)
DHOFAR Commissioned 1982
AL SHARQIYAH Commissioned 1983
Al BATNAH Commissioned 1984
Remarks: Vosper Thornycroft built all three craft, receiving the order for the first in 1980 and delivering on 21st October 1982. In the interim the other two craft were ordered. The Sultan of Oman's navy comprises 2,000 officers and men all volunteers, and 24 craft deployed at three bases, the most important of which is in the port of Muscat.

SPARVIERO CLASS
FAST ATTACK HYDROFOIL (MISSILE)

Country of Origin: Italy
Displacement: 62.5 tons
Dimensions: Length 24.6m (80.7ft) foilborne and 23.0m (75.4ft) for hull; beam 12.1m (39.7ft) foilborne

and 7.0m (22.9ft) for hull; draught 4.4m (14.4ft) hullborne and 1.6m (5.2ft) for hull

Gun Armament: One 76mm (3in) L/62 DP OTO-Melara Compact

Missile Armament: Two single container-launchers for two Otomat surface-to-surface missiles

Propulsion: CODOG

Performance: Maximum speed 50kts foilborne in calm sea or 8kts hullborne; range 2,225km (1,385 mls) at 8kts or 740km (460 mls) at 45kts

Complement: 2+8

Class

1. Italy (7)

SPARVIERO Commissioned July 1974

FEARLESS CLASS
ASSAULT SHIP

Country of Origin: UK
Displacement: 11,060 tons standard, 12,120 full load and 16,950 tons flooded down
Dimensions: Length 158.5m (520ft); beam 24.4m (80ft); draught 6.2m (20.5ft) normal and 9.8m (32ft) flooded down

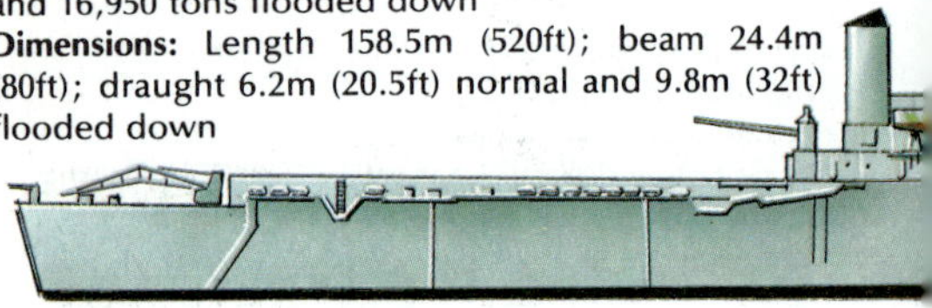

Gun Armament: Two 40mm (1.58in) Bofors L/70 AA in two single mountings
Missile Armament: Four quadruple launchers for Sea Cat surface-to-air missiles
Aircraft: Five Westland HU.Mk 5 helicopters
Capacity: The well deck can accommodate four LCM(9)s, which can be supplemented in the ship-to-shore role by four LCVPs carried in davits; typical internal load is 15 main battle tanks, seven 3-ton trucks and 20 Land Rovers, normal troop accommodation is 400, but up to 700 can be carried in austere conditions.
Propulsion: Two Babcock & Wilcox boilers supplying steam to two sets of English Electric geared turbines delivering 16,405kW (22,000shp) to two shafts
Performance: Maximum speed 21kts; range 9,250km (5,750 mls) at 20kts

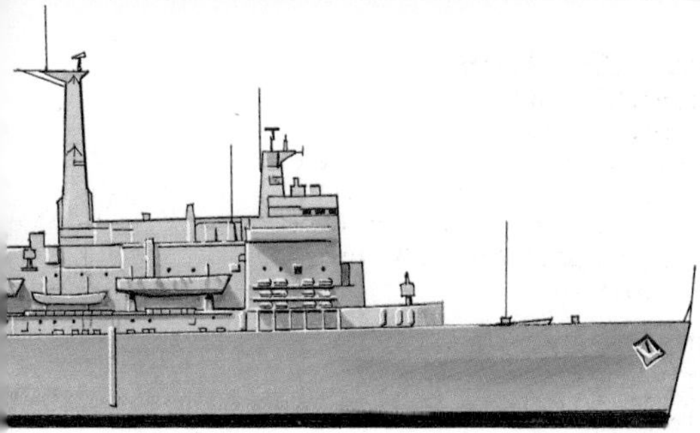

Complement: 580
Class
1. UK (2)
Name:
FEARLESS Commissioned November 1965
INTREPID Commissioned March 1967
Remarks: The two Fearless class assault ships were to have been scrapped during the 1980s until their vital role in the Falklands campaign of 1982 earned them a fully justified reprieve. Each ship is fitted out as the HQ unit for a naval assault group and landing brigade, though the capacity of each ship is somewhat short of a brigade of infantry. Though landing ships possess greater payload capability, the virtue of the assault ship lies with its speed, seakeeping and range, all factors of paramount significance in the Falklands campaign.

IWO JIMA CLASS
AMPHIBIOUS ASSAULT SHIP

Country of Origin: USA
Displacement: 18,000 tons full load

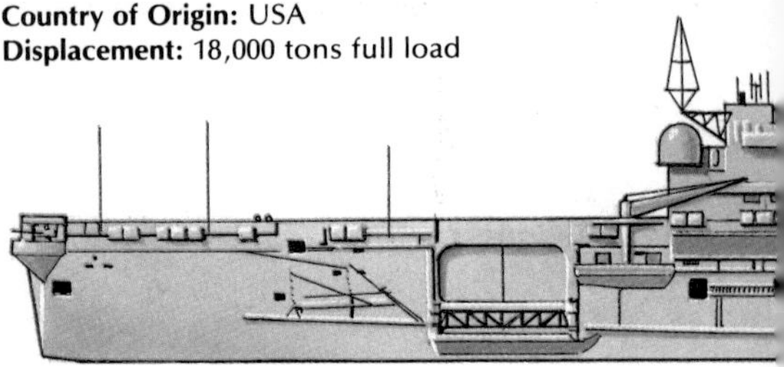

Dimensions: Length 183.7m (602.3ft); beam 25.6m (84ft); draught 7.9m (26ft); flight-deck width 31.7m (104ft)
Gun Armament: Four 76mm (3in) L/50 DP in two Mk 33 twin mountings, and (to be fitted) two 20mm (0.79in) Phalanx Mk 16 close-in weapon system mountings
Missile Armament: Two Mk 25 launchers for RIM-7 Sea Sparrow surface-to-air missiles
Aircraft: 20 Boeing Vertol CH-46 Sea Knight helicopters, or 11 Sikorsky CH-53 Sea Stallion helicopters, or a mixture of the two, and (with a reduction in helicopter strength) four BAe/McDonnell Douglas AV-8A Harrier aircraft
Capacity: Accommodation is provided for a US Marine Corps battalion landing team (144+1,602) plus its equipment, guns and vehicles; this accommodation amounts to 400m² (4,300 sq ft) for vehicles

and 3,475m^2 (37,400 sq ft) for palleted stores, plus bulk storage for 24,605 litres (6,500 US gal) of vehicle fuel and 1,533,090 litres (405,000 US gal) of helicopter fuel

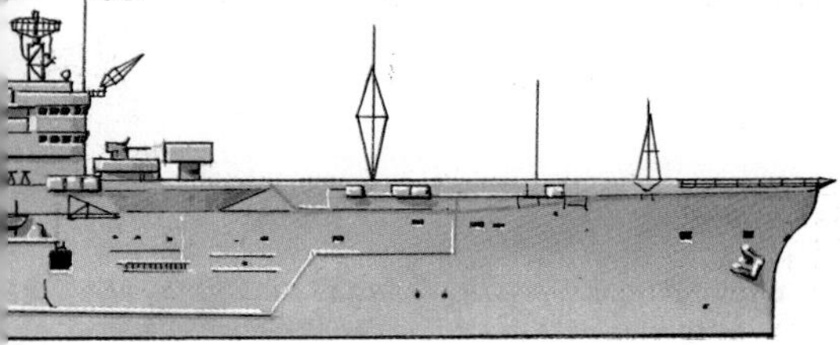

Propulsion: Two boilers (Babcock & Wilcox) supplying steam to one geared turbine delivering 16,405kW (22,000shp) to one shaft
Performance: Maximum speed 23kts
Complement: 47+562
Class
1. USA (7)
IWO JIMA Commissioned August 1961
Remarks: Of the seven Iwo Jima class amphibious assault ships, three are allocated to the Pacific and four to the Atlantic. They were the first ships in the world designed specifically for the delivery of an amphibious assault force by helicopter. Each ship can carry a US Marine Corps battalion landing team together with its weapons, vehicles and equipment, as well as a reinforced assault helicopter squadron.

TARAWA CLASS
AMPHIBIOUS ASSAULT SHIP

Country of Origin: USA

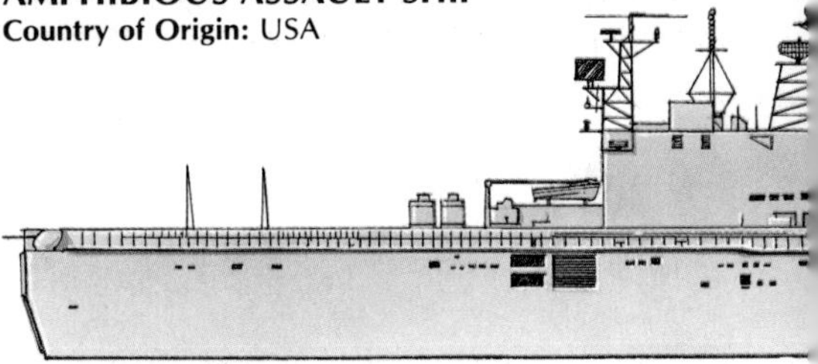

Displacement: 39,300 tons full load
Dimensions: Length 249.9m (820ft); beam 32.3m (106ft); draught 7.9m (26ft)
Gun Armament: Three 127mm (5in) L/54 DP in three Mk 45 single mountings, six 20mm (0.79in) AA in six Mk 67 single mountings, and (to be fitted) two 20mm (0.79in) Phalanx Mk 16 close-in weapon system mountings
Missile Armament: Two Mk 25 launchers for RIM-7 Sea Sparrow surface-to-air missiles (to be removed when CIWS is fitted)
Aircraft: Up to 19 Sikorsky CH-53 Sea Stallion or 26 Boeing Vertol CH-46 Sea Knight helicopters, and (by a reduction in embarked helicopter strength) a small number of BAe/McDonnell Douglas AV-8A Harrier aircraft
Capacity: Apart from two full length hangar decks under the flight deck, there is a docking well measur-

ing 81.7m (268ft) in length and 23.8m (78ft) in width to accommodate four LCU 1610 type landing craft, other landing capability being provided by six LCM(6) craft;

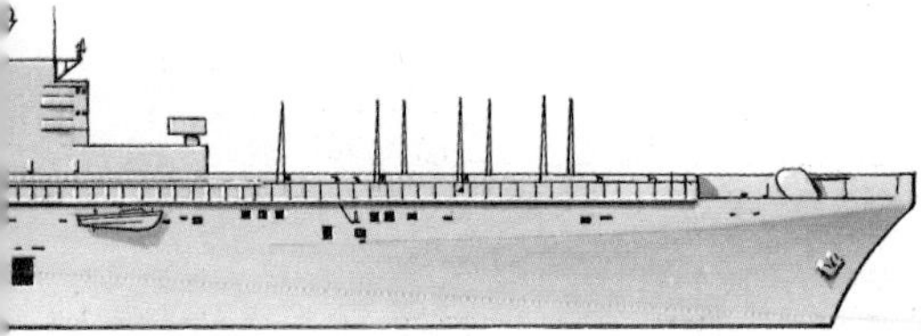

vehicle accommodation area is 3,135m² (33,730 sq ft), and stowage volume for palleted stores is 3,310m³ (116,900 cu ft); liquid storage is provided for 37,855 litres (10,000 US gal) of vehicle fuel and 1,514,160 litres (400,000 US gal) of helicopter fuel; troop accommodation is provided for a reinforced battalion of US Marines totalling 1,703 officers and men
Propulsion: Two Combustion Engineering boilers supplying steam to two sets of Westinghouse geared turbines delivering 104,400kW (140,000shp) to two shafts
Performance: Maximum speed 24kts; range 18,505km (11,500 mls) at 20kts
Complement: 90+812
Class
1. USA (5)
TARAWA Commissioned May 1976

IVAN ROGOV CLASS
LANDING SHIP DOCK
Country of Origin: USSR

Displacement: 14,000 tons full load
Dimensions: Length 159.0m (521.6ft); beam 24.5m (80.2ft); draught 6.5m (21.2ft)
Gun Armament: Two 76mm (3in) L/60 DP in one twin mounting, four 30mm (1.18in) AA 'Gatling' mountings, and one 40-barrel BM-21 rocket-launcher
Aircraft: Three to five Kamov Ka-25 'Hormone' helicopters on a platform forward and in a hangar aft
Capacity: The dock is some 76.0m (249.3ft) long, and the vessel is designed to carry a Naval Infantry Battalion of 522 men with up to 40 tanks; the docking well can accommodate two 'Lebed' class hovercraft and one 'Ondatra' class LCM, though this reduces tank capacity to 20
Electronics: One 'Head Net-C' 3D radar, one Owl

Screech' main armament fire-control radar, two 'Bass Tilt' AA fire-control radars, one 'Pop Group' SAM fire-control radar, one 'Don Kay' navigation radar, and one 'High Pole' IFF
Propulsion: Two gas turbines delivering 33,555kW (45,000shp) to two shafts
Performance: Maximum speed 26kts; range 18,500km

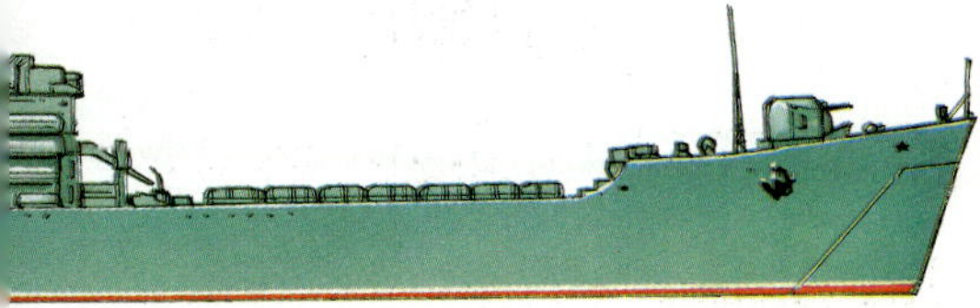

(11,500 mls) at 12kts
Complement: 400
Class
1. USSR (2)
IVAN ROGOV Commissioned 1978
ALEKSANDR NIKOLAEV Commissioned 1982
Remarks: The Ivan Rogov class heralds the emergence of a potent long-range amphibious assault capability for the USSR, each ship being able to transport a full battalion of naval infantry with its vehicles and other equipment. The bow ramp offers beaching capability, while the well makes offshore operations a simple matter. Good AA protection by SAMs and guns is provided.

AVENGER CLASS
MINE-COUNTERMEASURES VESSEL
Country of Origin: USA

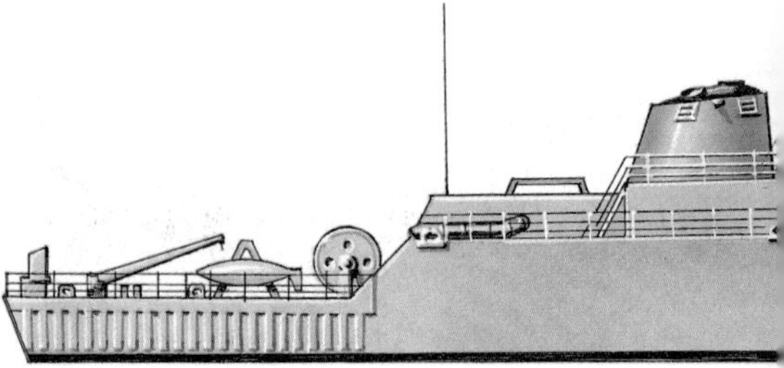

Displacement: 1,040 tons full load
Dimensions: Length 64.0m (210ft); beam 13.5m (44.3ft); draught 3.2m (10.5ft)
Gun Armament: Two 12.7mm (0.5in) heavy machine-guns
Electronics: One Westinghouse SPS-55 surface-search radar, and SQQ-30 sonar
Propulsion: Four Waukesha diesels delivering power to two shafts
Performance: Maximum speed 14kts
Complement: 5+57
Class

1. USA (0+1+20)
Name:
AVENGER: Commissioned 1985
Remarks: Being built to replace the Aggressive and Acme class, 21 ships of this new Avenger class have been ordered for building over the next ten years.

The first in the class *Avenger* was built by Peterson Builders Inc, Sturgeon Bay, Wisconsin and has a hull of oak and fir with a superstructure of fibreglass. A second class of mine-countermeasures vessel is planned to supplement the Avenger class, but details of this have not yet been announced.

CIRCE CLASS MINEHUNTER

Country of Origin: France

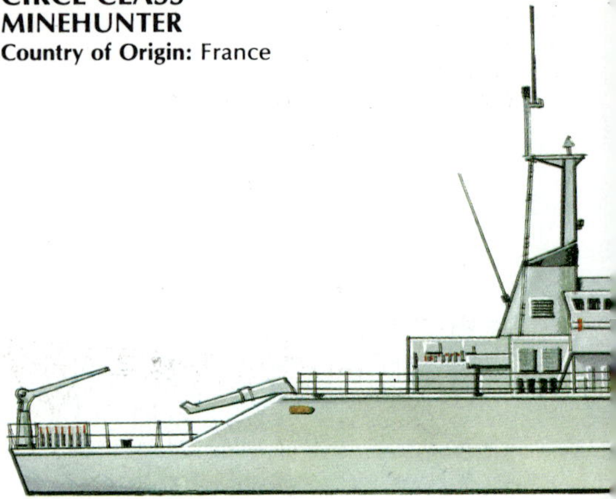

Displacement: 465 tons standard, 495 tons normal and 510 tons full load

Dimensions: Length 50.9m (167ft); beam 8.9m (29.2ft); draught 3.4m (11.15ft)

Gun Armament: One 20mm (0.79in) AA

Electronics: One navigation radar, one DUBM 20 minehunting sonar and two PAP wire-guided detonation sleds

Propulsion: One MTU diesel delivering 1,340kW (1,800bhp) to one shaft

Performance: Maximum speed 15kts; range 5,560km

(3,455 mls) at 12kts
Complement: 4+15+29
Class
1. France (5)
CYBELE Commissioned September 1972

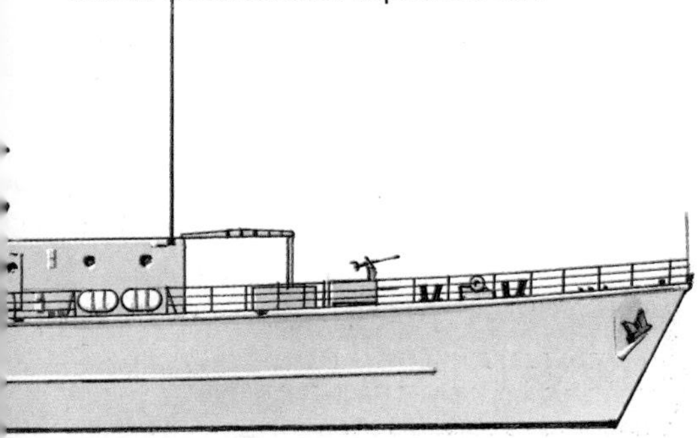

CALLIOPE Commissioned September 1972
CLIO Commissioned May 1972
CIRCE Commissioned May 1972
CERES Commissioned March 1973
Remarks: The *Cybele* (M 712) is built of laminated wood, and has two propulsion systems (one for conventional movement and the other for minehunting). The latter comprises two small and quiet propellers (at the base of the single main rudder) driven by 195-kW (260hp) electric motors for a speed of 7kts and extreme manoeuvreability.

NATYA I AND NATYA II CLASS OCEAN MINESWEEPERS

Country of Origin: USSR
Displacement: 650 tons standard and 950 tons full load

Dimensions: Length 61.0m (200.1ft); beam 10.0m (32.8ft); draught 3.5m (11.5ft)
Gun Armament: Four 30mm (1.18in) AA in two twin mountings, and four 25mm (1in) AA in two twin mountings
Anti-submarine Armament: Two RBU 1200 five-barrel rocket-launchers ('Natya I' only)
Electronics: One 'Drum Tilt' gun-control radar, one 'Don-2' navigation radar, one 'High Pole-B' IFF, and two 'Square Head' IFF
Propulsion: Two diesels delivering 5,965kW (8,000 shp) to two shafts
Performance: Maximum speed 20kts
Complement: 50

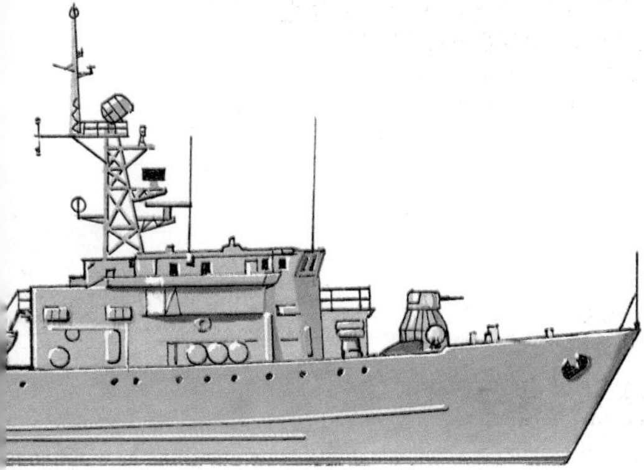

Class
1. India (6)
PONDICHERRY Commissioned 1978
2. Libya (4)
RAS EL GELAIS Commissioned 1981
3. USSR (34 Natya I and 1 Natya II Class)
Remarks: The Soviet Natya I class ocean minesweeper displays a basic similarity to other comparable vessels, with a large quarterdeck for the stowage and working of the sweeping gear. The 30mm (1.18in) mounts are on the centreline fore and aft, the 25mm (1in) mountings being staggered amidships.

HUNT CLASS
COASTAL MINESWEEPER/MINEHUNTER

Country of Origin: UK
Displacement: 615 tons standard and 725 tons full load
Dimensions: Length 60.0m (197ft); beam 10.0m (32.8ft); draught 2.5m (8.2ft)

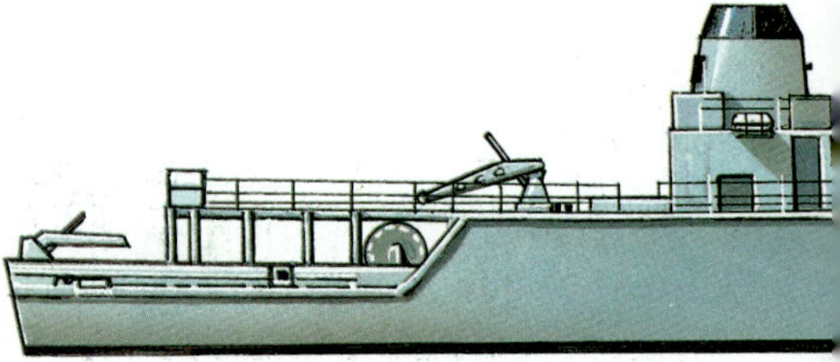

Gun Armament: One 40mm (1.58in) Bofors AA
Electronics: Navigation radar, and Type 2093 hull-mounted sonar
Propulsion: Two Ruston-Paxman 9-59K Deltic diesels delivering 2,835kw (3,800bhp) to two shafts
Performance: Maximum speed 16kts; range 2,775km (1,725 mls) at 12kts
Complement: 6+39
Class
1. UK (6+5)
Name:
BRECON Commissioned March 1980

LEDBURY Commissioned June 1981
CATTISTOCK Commissioned June 1982
COTTESMORE Commissioned March 1983

BROCKLESBY Commissioned February 1983
MIDDLETON Commissioned June 1984
DULVERTON Commissioned September 1983
CHIDDINGFOLD Commissioned August 1984
HURWORTH Commissioned 1986

Remarks: The Hunt class witnesses the Royal Navy's tardy realization of the obsolescence of the Ton class. The ships are undoubtedly capable, combining sweeping and hunting capability in an excellent GRP, glass-reinforced plastic (fibreglass) hull. Thorneycroft built seven with Yarrow providing the remaining two.

WILTON CLASS
COASTAL MINESWEEPER

Country of Origin: UK

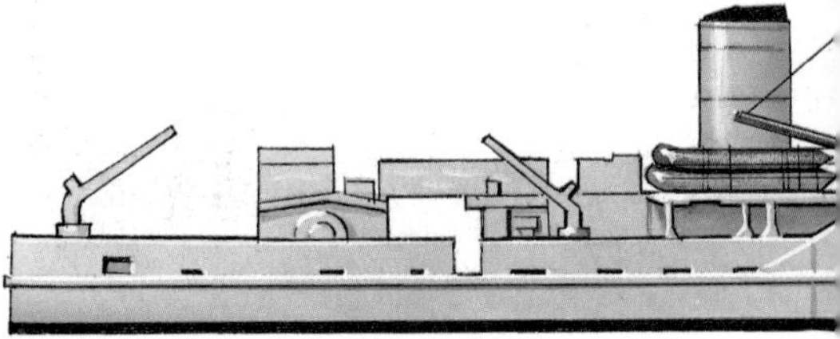

Displacement: 450 tons full load
Dimensions: Length 46.3m (153ft); beam 8.8m (28.8ft); draught 2.5m (8.5ft)
Gun Armament: One 40mm (1.58in) Bofors in Mk 7 single mounting
Electronics: One Kelvin Hughes Type 975 surface-search radar, one Type 955M radar, and one Plessey Type 193M hull-mounted minehunting sonar
Propulsion: Two English Electric Deltic 18-7A diesels delivering 2,235kW (3,000bhp) to two shafts
Performance: Maximum speed 16kts; range 4,265km

(2,650 mls) at 13kts
Complement: 5+32
Class
1.UK (1)
WILTON Commissioned July 1973
Remarks: The *Wilton* (M 1116) was the world's first GRP warship, and was ordered in 1970. The design is similar to that of the Ton class, and the ship is in fact fitted with reconditioned machinery from the scrapped Ton class *Derriton.*

SUBMARINES

In February 1956 the Americans announced that *Grayback* and *Growler*, originally laid down as diesel-electric attack submarines, were to be completed as guided-missile submarines for the deployment of Regulus cruise missiles. The two submarines were cut in two and cylindrical hangars installed above the bows to carry the missiles. Automatic loading and firing gear enabled the submarine to surface, extract a Regulus from the hangar straight onto an inclined built-in launch ramp, fire the missile and submerge

Grayback and *Growler* became operational in 1958 and within a year the Soviet Navy had countered with an equivalent stop-gap: the Whiskey Twin-Cylinder conversion. Much clumsier than the built-in hangars used in *Grayback* and *Growler*, the Whiskey Twin Cylinder was, as its name suggests, a Whiskey class boat with a couple of cylindrical missile hangars fitted aft of the sail. It gave the Soviets much-needed test experience with their new cruise missile, the SS-N-3 'Shaddock'. But it was, to quote *Jane's Fighting Ships*, 'probably never truly operational, being a thoroughly messy conversion which must make a noise like a train if proceeding at any speed above dead slow when dived'.

For the American submarine fleet, however, the *Grayback* and *Growler* conversions proved to be no more than stop-gaps. After a traumatic three years in which it had seemed that the Soviet Union was well on the way to establishing a decisive lead over the

United States in long-range rocketry, the Americans dramatically unveiled their Polaris programme and the giant new ships which would carry it.

The Polaris formula took the streamlined, teardrop submarine hull of the Skipjack class – and stretched it by a breathtaking 130 feet, from 251 to 381 feet in overall length. The huge hull volume thus created was used to accommodate two rows of eight vertical rocket-launching tubes abaft the fin. Each of these 16 missiles could be assigned a different target and all were designed for underwater launching, the rocket engine cutting in as the missile broke surface and the guidance system correcting the missile's attitude to provide the correct flight trajectory to the target. The real versatility of the Polaris formula was – and remains – immunity from retaliation by enemy land-based missiles, plus the unlimited initial radius provided by the submarine's nuclear power plant before a single missile is even fired.

The decision to introduce Polaris was taken while the United States was still being rocked by Sputnik mania in the winter of 1957-58. Three new submarines specifically designed for carrying Polaris were ordered in the Supplemental New Construction Programme signed on 11 February 1958. Two more were ordered in July 1958. The lead ship of the class, *George Washington,* was launched on 9 June 1959 and commissioned on 15 November of that year. On 20 July 1960, lying submerged off the Florida coast, she made the first successful firing of the new weapon. Two missiles were launched, their performance being carefully monitored from the nearby 'spaceport' on Cape Canaveral. The test was successful: the Polaris era had arrived. Two years later the Polaris system was given its first complete test, from underwater firing to warhead detonation. This happened on 6 May 1962 when *Ethan Allen,* first of the second batch of five Polaris submarines laid down between 1959 and 1961, fired a Polaris A-2 in the Pacific Test Area of Christmas Island.

With the Polaris concept, the role of the submarine as an instrument of undersea war changed completely. There were no precedents for this change, nor any convenient parallel case histories of land, air or sea weaponry to make the change any easier to describe. The 'Strategic Ballistic Missile Submarine' or SSBN did not render all other types of nuclear and diesel-electric submarines obsolete. On the contrary, these continue to be built and introduced to service as a vital element of conventional naval deployment, supplementary to the posture of strategic deterrence. But the Polaris SSBN was the first warship deliberately designed to threaten the destruction of 16 enemy cities with their civilian populations. The grim novelty of this role makes it impossible to consider the SSBN in the context of conventional undersea war.

The advent of Polaris and its manifest advantages caused extensive restructuring of nuclear deterrent forces, both inside the United States and among her

NATO allies. As more and more American SSBNs came into service through the 1960s and the stockpile of nuclear warheads grew, the former importance of strategic bombing forces began to recede. For example, the total of American strategic bombers dropped from 630 in 1966 to 373 in 1977. After a costly flirtation with the American Skybolt stand-off bomb, Britain in 1962 decided to wind down her strategic bombing force and adopt Polaris herself, building her own submarines but taking the American missile. The four ships of the Resolution class were laid down in 1964-65; *Resolution*, the first into service, was commissioned in October 1967. France followed suit with her Redoutable class, the first of which, *Redoutable*, became operational in December 1971. Unlike Britain, France elected to develop her own missile 'family'. (Considerably advanced in rocket development over her European neighbours, the French joined the ranks of the space powers by putting their first satellite into orbit in November 1965, launched from a French-made Diamant rocket.) France went on to fit her SSBN fleet with her own ballistic missile, the MSBS.

In retrospect, one of the most intriguing points about the adoption of the SSBN is that it took place during the last 15 years of world affluence, 1958-73, before the Yom Kippur War in the Middle East and the ensuing world energy crisis. *Skipjack* had cost $40 million; *Ethan Allen* cost $105 million. As for the main American SSBN fleet of no less than 31 ships

(Lafayette and Benjamin Franklin classes, laid down between January 1961 and March 1965) they cost about $109.5 million each. However, these were only the construction costs; on top of the amounts mentioned above one must add the cost of replacing nuclear cores (about $3.5 million apiece) and repeatedly up-dating the ballistic missiles as bigger and better models are developed.

Having elected to join the SSBN nuclear deterrers, however, Britain could certainly not afford to keep pace with the high-speed development of SLBM missiles. The original Polaris A-1 missile had a range of 2,145km (1,333mls). This was almost immediately improved by Polaris A-2 with 2,683km (1,666mls), and in October 1963 the *Andrew Jackson* fired the first Polaris A-3 off Cape Canaveral. Polaris A-3 not only had a vastly improved range of 4,470km (2,777mls) but it also delivered a cluster of three separate nuclear warheads to the same target. This 'parcelling' of nuclear warheads ushered in the second generation of submarine-launched ballistic missiles ponderously initialled as MIRV; 'Multiple Independently targetable Re-entry Vehicle'. In the case of Poseidon, the first SLBM MIRVed missile and successor to Polaris A-3, this mèant a missile which, like Polaris, can be carried by an SSBN and launched from under water with a range of 4,470km (2,777mls). But once in flight its payload splits into ten mini-missiles, each capable of being steered to a separate target. Thus instead of being able to hit 16 enemy targets, as with

Polaris, an SSBN armed with Poseidon can hit 160. The ten George Washingtons and Ethan Allens of the SSBN 'first generation' were not suitable for conversion to Poseidon, but by 1978 all 31 of the Lafayettes and Benjamin Franklins had been fitted with Poseidon. With the Polaris missiles of the earlier SSBNs, the overall American SSBN fleet was now able to fire a combined theoretical total (theoretical, because not all the ships are simultaneously at sea) of 5020 nuclear warheads – 55 percent of the total United States nuclear deterrent.

In 1976 the Americans laid down the first of the giant (18,000 tons) Ohio class SSBNs. *Ohio,* the first of the class, was launched in November 1978. She carries no less than 24 launching tubes for the 'third-generation' SLBM, Trident, with MIRVed warhead and a nominal range of 7,155km (4,444mls). It was hard to see the SSBN being evolved much further until, in mid-November 1980, the Soviet Union launched the first of her Typhoon class giants. At 30,000 tons they are far bigger than many NATO aircraft carriers, and though their 20 SS-N-18 SLBMs are believed not to be MIRVed, they out-range Trident by 2,145km (1,333mls). Caught completely on the wrong foot by the early American commitment to the SSBN, the Soviet Union spent a desperate decade of improvisation and experiment to develop an equivalent system. The first Russian method used for firing missiles from submarines (apart from cruise missiles employed in the long-range SSM role) consisted of lengthening the sail to produce a streamlined fairing capable of carrying ballistic missiles in two or three vertically mounted tubes. This process, begun as early as 1955 with converted Aulu class boats, actually made the Soviet Union the first sea power to produce a ballistic missile submarine. But the only Russian missile small enough to be carried in a submarine was the tiny 'Sark' with a range of no more than 483km (300mls), and by the late 1950s it was not to be expected that any diesel-electric submarine would be able to get within 483 km of the American continent without being detected.

Nevertheless, this was the system with which the Russians proceeded into the 1960s, building small successive batches of submarines capable of carrying missiles of increasing range. The Golf variants, 21 in all, were accompanied by the 13 Hotel class, appearing between 1958 and 1962; both were originally equipped with 'Sark', but subsequently equipped with the SS-N-5 'Serb', with a range of 1,250km (777mls). Experience thus gained led to the USSR's answer to the American Polaris SSBN: the 8500-ton Yankee class, which appeared in 1967. The Yankee SSBN imitated the whale-shaped, elongated hull of the American boats and also mounted its missiles in 16 vertical tubes – but the range of its SSN-6 Sawfly missiles was only 2,325km (1,444mls), and even when improved in later marks this range still fell far short of the 4,470km (2,777mls) of Polaris A-3 and Poseidon.

The 1970s have seen the Soviet Union doggedly

catching up with the enormous lead established by the American SSBN force in the 1960s. The Delta class SS-N-18 missile took the Soviet SSBN force through the 8,000km (5,000mls) range barrier, and the first Soviet experiments with MIRVing were soon detected. In 1976, the first year in which comparisons in MIRVed warheads could be drawn, the Soviet Union was estimated to have only 140 missiles to the American 1046 – or 2970 independently targetable warheads (ICBMs and SLBMs combined) to the American 7274. By 1976 the Russians had produced the Delta III, armed with the SS-N-18 of no less than 8,000km (5,000mls) range – capable of hitting any part of the northern hemisphere from Soviet-controlled waters. Soviet MIRVing continues to improve and the likelihood of the Soviet Union achieving MIRV parity in the near future can be no more than delayed by the American Trident programme. The newest, 30,000-ton Typhoons are also designed for the SS-N-18, but will almost certainly be used as test-beds for more sophisticated MIRVed Soviet SLBMs in the near future.

OBERON AND PORPOISE CLASS
PATROL SUBMARINE

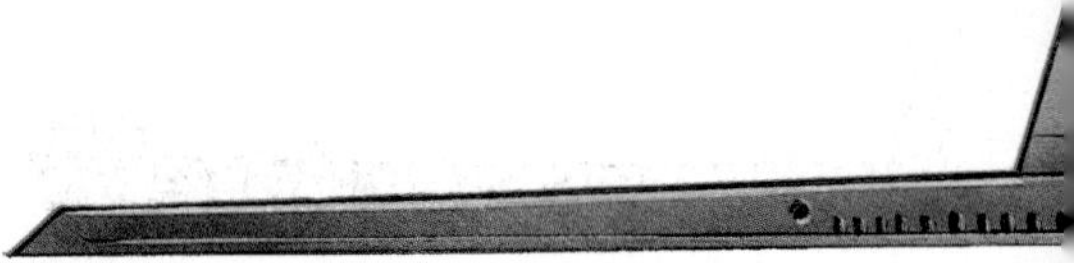

Country of Origin: UK
Displacement: 1,610 tons standard, 2.030 tons surfaced and 2,410 tons dived
Dimensions: Length 90.0m (295.2ft); beam 8.1m (26.5ft); draught 5.5m (18ft)
Torpedo Armament: Six 533mm (21in) tubes (all bow)
Anti-submarine Armament: Two 533mm (21in) tubes (both stern) for short A/S torpedoes carried as part of the total of 24 torpedoes
Propulsion: Diesel-electric arrangement, with two Admiralty Standard Range diesels delivering 2,745kW (3,680bhp) and two electric motors delivering 4,475kW (6,000shp) to two shafts
Performance: Maximum speed 12kts surfaced and 17kts submerged; range 16,655km (10,350 mls) surfaced at cruising speed; diving depth 275m (900ft)
Complement: 7+62

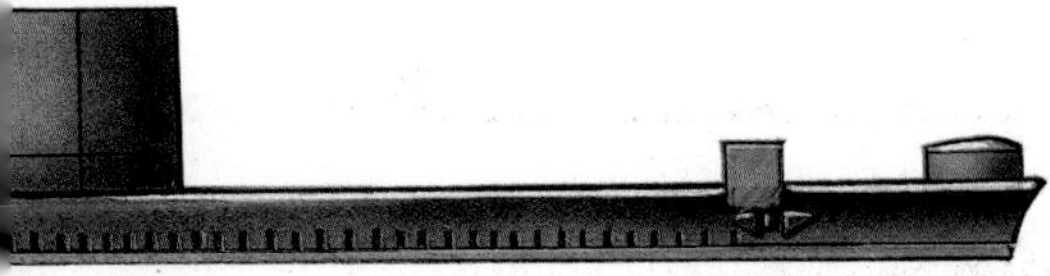

Remarks: The Australian boats, alternatively designated RAN Oberon class, were ordered as four boats in 1963 and two additional units in 1971. They are similar in most respects to the British boats, but are currently being upgraded with provision for Mk 48 torpedoes and tube-launched Sub-Harpoon anti-ship missiles. The Brazilian boats were ordered in 1969 (two units) and 1972 (one unit). They are similar to the British Oberons but have the Vickers TIOS-B fire-control system. The Canadian boats are also similar to the British boats, but have more capable air-conditioning equipment, and Canadian operational equipment and communications gear. As part of the SOUP (Submarine Operational Update Project) modernization, the Canadian boats are being fitted with the Sperry Micropuffs passive ranging sonar and the Singer Librascope fire-control system. On 2nd December 1986 Britain's largest submarine 'Upholder' was launched at Barrow-in-Furness the first of the type 2400 diesel electrics.

ECHO II CLASS
NUCLEAR-POWERED CRUISE-MISSILE SUBMARINE

Country of Origin: USSR
Displacement: 4,800 tons surfaced and 5,800 tons dived
Dimensions: Length 117.3m (384.7ft); beam 9.2m (30.2ft); draught 7.8m (25.5ft)
Missile Armament: Eight launch tubes for SS-N-3A 'Shaddock' surface-to-surface missiles, or for SS-N-12 'Sandbox' surface-to-surface missiles
Torpedo Armament: Six 533mm (21in) tubes (all bow) for 24 torpedoes
Anti-submarine Armament: Two 406mm (16in) tubes (both stern) for two A/S torpedoes, plus 16 A/S torpedoes in total above
Propulsion: Two nuclear reactors supplying steam to two sets of geared turbines delivering 22,370kW (30,000shp) to two shafts
Performance: Maximum speed 25kts submerged; diving depth 300m (985ft) operational and 500m (1,640ft) maximum
Complement: 100
Remarks: The 'Echo II' class SSGN carries eight angled tubes for SS-N-3A (23 boats) or SS-N-12 (six boats) cruise missiles, and was built in the first half of the 1960s for attacks on US carrier forces.

SS-N-3A SHADDOCK

Country of Origin: USSR

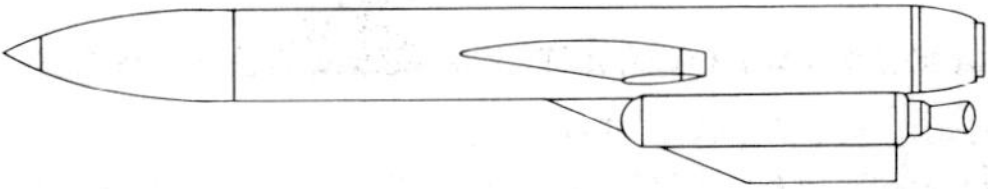

Type: Submarine-launched (surface) medium/long-range anti-ship cruise missile

Dimensions: Length about 13.0m (42ft 7.8in); span about 2.1m (6ft 10.7in); diameter about 1.0m (3ft 3.4in)

Weights: Total round about 4,500kg (9,921lb); warhead about 1,350kg (2,976lb) high explosive or 350 kiloton nuclear

Guidance: Probably radio command (or autopilot) with mid-course correction and active radar seeker in terminal phase

Control: Aerodynamic controls on the flying surfaces

Propulsion: For boost two jettisonable solid-propellant rockets, and for sustain one integral turbojet or ramjet

Performance: Maximum speed about Mach 2.2; range up to 850km (528mls)

Status: Operational since 1960, the SS-N-3 is still a powerful weapon with great capability against major warships thanks to the use of a large HE warhead or against task forces with the alternative nuclear warhead.

SWIFTSURE CLASS
NUCLEAR-POWERED FLEET SUBMARINE

Country of Origin: UK
Displacement: 4,200 tons standard and 4,500 tons dived
Dimensions: Length 82.9m (272ft); beam 9.8m (32.3ft); draught 8.2m (27ft)
Missile Armament: Five Sub-Harpoon underwater-to-surface missiles
Torpedo Armament: Five 533mm (21in) tubes (all bow) for 20 torpedoes
Anti-submarine Armament: A/S torpedoes included in the total above
Propulsion: One pressurized-water cooled Rolls-Royce nuclear reactor supplying steam to one set of General Electric geared turbines delivering 11,185kW (15,000shp) to one shaft; there is also an auxiliary Paxman diesel delivering 2,985kW (4,000hp)
Performance: Maximum speed 28kts surfaced and 30+kts submerged; diving depth 400m (1,325ft) operational and 600m (1,985ft) maximum
Complement: 12+85
Remarks: The Swiftsure class is the Royal Navy's second-generation SSN, with a larger hull than the Valiant class for greater internal volume and deeper diving combined with higher speed. Each of the five torpedo tubes can be reloaded in only 15 seconds.

LE REDOUTABLE CLASS
NUCLEAR-POWERED BALLISTIC-MISSILE SUBMARINE

Country of Origin: France
Displacement: 8,045 tons surfaced and 8,940 tons dived
Dimensions: Length 128.7m (422.1ft); beam 10.6m (34.8ft); draught 10.0m (32.8ft)
Missile Armament: Launch tubes for 16 MSBS M-20 underwater-to-surface ballistic missiles
Torpedo Armament: Four 533mm (21in) tubes for 18 torpedoes
Anti-submarine Armament: A/S torpedoes carried as part of the total above
Propulsion: One pressurized-water cooled nuclear reactor supplying steam to two sets of geared turbines driving two turbo-alternators providing current to one electric motor delivering 11,930kW (16,000hp) to one shaft; there is also an auxiliary powerplant comprising two diesels delivering 975kW (1,307hp)
Performance: Maximum speed 20+kts surfaced and 25kts submerged; range 9,250km (5,750 mls) on auxiliary diesels; diving depth more than 300m (985ft)
Complement: 15+120

LAFAYETTE CLASS
NUCLEAR-POWERED BALLISTIC-MISSILE SUBMARINE

Country of Origin: USA
Displacement: 6,650 tons light, 7,250 tons surfaced and 8,250 tons dived
Dimensions: Length 129.5m (425ft); beam 10.1m (33ft); draught 9.6m (31.5ft)
Missile Armament: Vertical launch tubes for 16 submarine-launched UGM-73A Poseidon C3 ballistic missiles
Torpedo Armament: Four 533mm (21in) Mk 65 tubes (all bow)
Anti-submarine Armament: UUM-44A SUBROC tube-launched missiles
Electronics: SSBN sonar, Mk 84 tactical computer, three Mk 2 Ship's Inertial Navigation System units, Mk 113 torpedo fire-control system, WSC-3 satellite communications transceiver and a navigational satellite receiver
Propulsion: One Westinghouse S5W pressurized-water cooled reactor supplying steam to two geared turbines delivering 11,185kW (15,000shp) to one shaft
Performance: Maximum speed 20kts surfaced and about 30kts submerged; diving depth more than 305m (1,000ft)

Complement: 20+148
Class
1. USA (19)
LAFAYETTE Commissioned April 1963
Remarks: The Lafayette and basically identical Benjamin Franklin class SSBNs were the largest underwater craft built by the Western alliance during the 1960s, and still constitute the most formidable portion of its submarine-launched missile deterrent capability pending the arrival of the vast Ohio class boats. Each boat is assigned two crews, one manning the boat during a 70-day patrol and then helping the alternate crew during the 32-day refit before the next patrol, which is manned by the second crew. Every six years the boats each undergo a major refit lasting just under two years. The nuclear cores fitted in the late 1960s and early 1970s provide energy for about 643,720km (400,000 miles), and the Lafayette class boats currently fitted with the Trident IC4 missiles are SSBNs 627, 629, 630, 632, 633 and 634. All the boats of the Lafayette and Benjamin Franklin classes are currently operational with the Atlantic Fleet.

RESOLUTION CLASS
NUCLEAR-POWERED BALLISTIC-MISSILE SUBMARINE

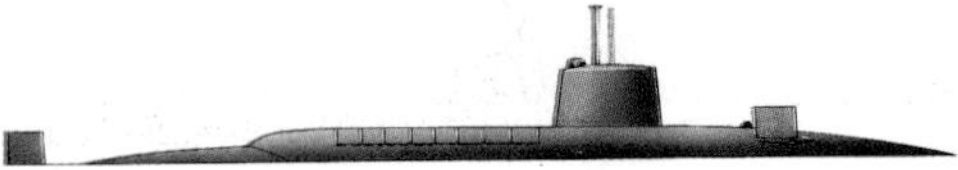

Country of Origin: UK
Displacement: 7,500 tons surfaced and 8,400 tons dived
Dimensions: Length 129.5m (425ft); beam 10.1m (33ft); draught 9.1m (30ft)
Missile Armament: Launch tubes for 16 UGM-27C Polaris A-3 underwater-to-surface ballistic missiles
Torpedo Armament: Six 533mm (21in) tubes (all bow)
Anti-submarine Armament: A/S torpedoes fired from the tubes above
Propulsion: One pressurized-water cooled Rolls-Royce nuclear reactor supplying steam to one set of English Electric geared turbines delivering 11,185kW (15,000shp) to one shaft
Performance: Maximum speed 20kts surfaced and 25kts submerged
Complement: 13+130
Remarks: In common with other navies' SSBNs, the Resolution class boats have two crews for maximum patrol time. The class is due to be replaced in the 1990s by a new class with Trident ballistic missiles.

TYPHOON CLASS
NUCLEAR-POWERED BALLISTIC-MISSILE SUBMARINE

Country of Origin: USSR
Displacement: About 30,000 tons dived
Dimensions: Length 183.0m (600.4ft), beam 22.9m (75.1ft); draught 15m (49.2ft)
Missile Armament: Launch tubes for 20 SS-N-20 underwater-to-surface ballistic missiles
Torpedo Armament: Six or eight 533mm (21in) tubes
Anti-submarine Armament: A/S torpedoes
Electronics: Surface-search radar, navigation radar, sonar, inertial navigation systems, missile fire-control system, and torpedo fire-control system
Propulsion: Two nuclear reactors supplying steam to two sets of geared turbines delivering about 89,500kW (120,020shp) to two shafts
Performance: Maximum speed about 24kts submerged
Complement: About 150
Remarks: The Typhoon class SSBNs are the largest submarines ever built. The construction features twin outer hulls, and the forward-located launch tubes leave the after section free for a twin-reactor powerplant.

LOS ANGELES CLASS
NUCLEAR-POWERED FLEET SUBMARINE

Country of Origin: USA

Displacement: 6,000 tons standard and 6,900 tons dived

Dimensions: Length 109.7m (360ft); beam 10.1m (33ft); draught 9.9m (32.2ft)

Missile Armament: Tube-launched RGM-84A Harpoon underwater-to-surface missiles, and BGM-109 Tomahawk underwater-to-surface missiles

Torpedo Armament: Four 533mm (21in) tubes (all amidships) for 26 weapons

Anti-submarine Armament: Mk 48 A/S torpedoes and UUM-44 SUBROC missiles

Propulsion: One pressurized-water cooled General Electric S6G nuclear reactor supplying steam to two sets of geared turbines delivering about 26,100kW (35,000shp) to one shaft

Performance: Maximum speed 31kts submerged; diving depth 450m (1.475ft) operational and 750m (2,460ft) maximum

Complement: (12+115)

Remarks: The Los Angeles class of fleet submarines is planned as the US Navy's standard attack boat into the 21st century, and marks a considerable advance in underwater capability by combining the weapons and sensors of the Thresher and Sturgeon classes with a more refined hull form and greatly improved powerplant. If the Los Angeles class boats are the only US Navy SSNs able to match the speed of carrier task groups, allowing them to be deployed with such groups as counters to Soviet SSNs and SSGNs. The sensor fit is admirable, but the first 33 boats are having their complement of ASW torpedoes cut to permit the carriage of 12 tube-launched BGM-109 Tomahawk cruise missiles with anti-ship warheads. From the 34th boat onward 15 vertical launch tubes for Tomahawks will be fitted between the pressure and outer hulls in the area of the sonar dome in the bows, returning ASW weapon strength (torpedoes and SUBROC missiles) to 26.

DELTA I CLASS
NUCLEAR-POWERED BALLISTIC-MISSILE SUBMARINE

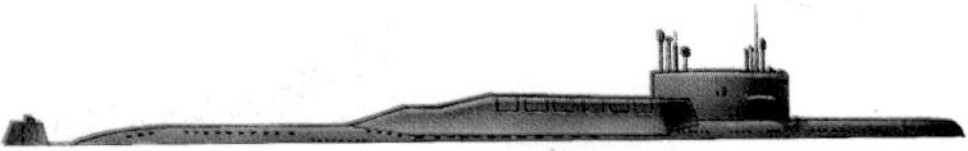

Country of Origin: USSR
Displacement: 10,000 tons dived
Dimensions: Length 136.0m (446.1ft); beam 11.6m (38ft); draught 10.0m (32.8ft)
Missile Armament: 12 launch tubes for SS-N-8 underwater-to-surface ballistic missiles
Torpedo Armament: Six 533mm (21in) tubes (all bow) for 18 torpedoes
Anti-submarine Armament: A/S torpedoes
Propulsion: Two nuclear reactors supplying steam to two sets of turbines delivering 22,370kW (30,000shp) to two shafts
Performance: Maximum speed 25kts submerged
Complement: 120
Remarks: Built from 1972, the Delta I class could carry only 12 instead of 16 SLBMs because of the SS-N-8's greater length and diameter. Built at Severodvinsk and Komosomolsk, the Delta I class SSBNs superseded the Yankee class, the greater length of the SS-N-8 missiles requiring a heightened casing on the after section carrying the 12 launch tubes.

VICTOR I CLASS
NUCLEAR-POWERED FLEET SUBMARINE

Country of Origin: USSR
Displacement: 4,200 tons surfaced and 5,200 tons dived
Dimensions: Length 93.9m (307.7ft); beam 10.0m (32.8ft); draught 7.3m (23.9ft)
Torpedo Armament: Six 533mm (21in) tubes (all bow) for 18 torpedoes
Anti-submarine Armament: 10 A/S torpedoes, and two SS-N-15 underwater-to-underwater missiles
Propulsion: Two nuclear reactors supplying steam to two sets of geared turbines delivering 22,370kW (30,000shp) to one main and two auxiliary shafts
Performance: maximum speed 20kts surfaced and 32kts submerged; diving depth 400m (1,325ft) operational and 600m (1,970ft) maximum
Complement: 90
Remarks: The Victor classes still form the main strength of the Soviet fleet submarine strength. The Victor I appeared in 1967 and introduced teardrop shape to the Soviet navy, Victor II in 1972 and Victor III in 1976.

Index